50 Years of Engineering
in Singapore

World Scientific Series on Singapore's 50 Years of Nation-Building

---

Published

50 Years of Social Issues in Singapore
   edited by David Chan

Our Lives to Live: Putting a Woman's Face to Change in Singapore
   edited by Kanwaljit Soin and Margaret Thomas

50 Years of Singapore–Europe Relations: Celebrating Singapore's Connections with Europe
   edited by Yeo Lay Hwee and Barnard Turner

Perspectives on the Security of Singapore: The First 50 Years
   edited by Barry Desker and Cheng Guan Ang

50 Years of Singapore and the United Nations
   edited by Tommy Koh, Li Lin Chang and Joanna Koh

50 Years of Environment: Singapore's Journey Towards Environmental Sustainability
   edited by Tan Yong Soon

Food, Foodways and Foodscapes: Culture, Community and Consumption in Post-Colonial Singapore
   edited by Lily Kong and Vineeta Sinha

50 Years of the Chinese Community in Singapore
   edited by Pang Cheng Lian

Singapore's Health Care System: What 50 Years Have Achieved
   edited by Chien Earn Lee and K. Satku

Singapore–China Relations: 50 Years
   edited by Zheng Yongnian and Lye Liang Fook

Singapore's Economic Development: Retrospection and Reflections
   edited by Linda Y. C. Lim

Singapore and UNICEF: Working for Children
   edited by Peggy Kek and Penny Whitworth

Singapore's Real Estate: 50 Years of Transformation
   edited by Ngee Huat Seek, Tien Foo Sing and Shi Ming Yu

The Singapore Research Story
   edited by Hang Chang Chieh, Low Teck Seng and Raj Thampuran

The complete list of titles in the series can be found at
http://www.worldscientific.com/series/wss50ynb

World Scientific Series on
Singapore's 50 Years of Nation-Building

50 Years of Engineering in Singapore

Editor

Cham Tao Soon
President (Emeritus), Nanyang Technological University

Contributions by
The Institution of Engineers, Singapore

NEW JERSEY • LONDON • SINGAPORE • BEIJING • SHANGHAI • HONG KONG • TAIPEI • CHENNAI • TOKYO

Published by

World Scientific Publishing Co. Pte. Ltd.
5 Toh Tuck Link, Singapore 596224
USA office: 27 Warren Street, Suite 401-402, Hackensack, NJ 07601
UK office: 57 Shelton Street, Covent Garden, London WC2H 9HE

British Library Cataloguing-in-Publication Data
A catalogue record for this book is available from the British Library.

World Scientific Series on Singapore's 50 Years of Nation-Building
50 YEARS OF ENGINEERING IN SINGAPORE

Copyright © 2018 by World Scientific Publishing Co. Pte. Ltd.

All rights reserved. This book, or parts thereof, may not be reproduced in any form or by any means, electronic or mechanical, including photocopying, recording or any information storage and retrieval system now known or to be invented, without written permission from the publisher.

For photocopying of material in this volume, please pay a copying fee through the Copyright Clearance Center, Inc., 222 Rosewood Drive, Danvers, MA 01923, USA. In this case permission to photocopy is not required from the publisher.

ISBN 978-981-4632-28-7
ISBN 978-981-4632-29-4 (pbk)

Typeset by Stallion Press
Email: enquiries@stallionpress.com

Foreword

Engineering has been the cornerstone of Singapore's astonishing 50-year economic, infrastructural and societal transformation. The tireless labour and professional excellence of our engineers not only helped our young nation overcome existential challenges in our early days of independence, but also created good jobs, wealth and quality living for our people.

Thanks to the work of engineers, Singapore has enjoyed robust growth in key industry clusters such as electronics, manufacturing, construction and related services. Our standing as a global hub in manufacturing, logistics, finance and healthcare has also been enhanced by our engineering advancements.

As Singapore charts its future growth, the role of engineers has become more critical. Continued excellence in engineering will enable us to achieve our national objective of becoming a Smart Nation and a Sustainable and Liveable City; help address challenges such as climate change, manpower and productivity, and land-use; and create new opportunities through disruptive technologies.

I therefore welcome the publishing of *50 Years of Engineering in Singapore* by The Institution of Engineers, Singapore in collaboration with World Scientific Publishing. This provides our younger generation with an insight into the fascinating journeys of our pioneer engineers, and inspires them to embark on their own exciting and rewarding journeys unique to the engineering profession.

I would like to thank all our engineers who have contributed to Singapore's development. Your determination, wisdom and ingenuity have built Singapore into a lovely home for all of us.

I hope that this publication will excite our younger generation to join the ranks of the engineering profession, to build an even better future for all of us.

Mr Teo Chee Hean
Deputy Prime Minister
& Coordinating Minister for National Security
Singapore

Foreword

Engineering has been central to the well-being and economic development of Singapore since its founding in 1965. The work of engineers has enabled the nation's metamorphic transformation into a world-class financial hub and urban city state with a first-rate living environment.

Pioneering a future of change, many engineering forerunners had toiled industriously to make long-lasting impact to Singapore's first 50 years of growth. But the magnitude of their contributions has somewhat been masked by the typically humble inclination of engineers to stay behind the scenes.

To ensure that our current and future generations are well aware of how engineering has built Singapore from the ground up, The Institution of Engineers, Singapore (IES) is pleased to publish *50 Years of Engineering in Singapore*, or *50 YES* in collaboration with World Scientific Publishing and with Professor Cham Tao Soon as its editor.

Our hope is that by spotlighting the transformational contributions of engineers in our past, the nation will develop a greater appreciation for engineering and its pivotal role in our future success.

50 YES is far from being a laborious technical read. Written for the young and the public, *50 YES* is an exciting compendium packed with insights into multiple sectors of engineering. Delving beneath the surface, the book will give unique peeks into the remarkable thought processes and creative journeys of engineers. It will also feature bright young engineering talents representative of Singapore's future with the hope of inspiring many more to join this exciting profession. IES will be making the book available at major bookstores and online.

50 YES also marks IES' 50-year journey since our founding on 1 July 1966, as the national society of engineers of Singapore. Over the past five decades, IES has stayed true to its role of being the heart and voice of engineers, continuously striving to raise the professional competence and status of engineers. Moving into its 51st year, IES will build on the foundation laid down by pioneering engineers to support future advancement of engineering.

IES would like to extend our deepest appreciation to Deputy Prime Minister Teo Chee Hean for his support of this book and of our endeavour to promote engineering. We would also like to thank Professor Cham for kindly taking up the role of editor of the book and all our partners who have contributed to the content and production of this book.

50 YES is for every Singaporean to relish our rich legacy in engineering and to inspire more extraordinary ones in the future. Enjoy the read.

Er. Edwin Khew Teck Fook PBM
President
The Institution of Engineers, Singapore (IES)

Editor's Remark

It was Dr Phua Kok Khoo who approached me to create a volume on engineering achievements in Singapore in the last 50 years. This is to celebrate the 50th anniversary of the founding of Singapore as a nation. I agreed immediately because I believe the rapid development of Singapore was largely due to the contribution by engineers and engineering enterprises both in the private and public sectors. Engineers are known worldwide as the creators of wealth.

I knew the Institution of Engineers, Singapore (IES) would be the most appropriate organisation to write the 400 pages of engineering achievements. IES has a wide range of memberships both junior and senior and coming from all sectors. The Council members are all volunteers and dedicated to the promotion of engineering. I knew I could depend on them to do a good job. My past experience as a past President of the Council gave me the confidence.

I recall I had a successful brainstorming session with the 2015 Council of IES. We mapped out the engineering sectors to be highlighted and the appropriate groupings. I left the details of organising the compilations to the committees which had been created.

I have to thank all the writers for the hard work. I am confident all the contributors and IES would be very proud of what they have achieved. It is an authoritative document on the achievements in engineering in Singapore for the last 50 years.

Prof Cham Tao Soon
PhD(Cambridge), FREng, FSEng
President (Emeritus)
Nanyang Technological University
Singapore

Acknowledgements

The Editor would like to thank the following contributors for their input:

Dr Bow Jaw Woei, *Fellow, The Institution of Engineers, Singapore*
Dr Chew Soon Hoe, *Assistant Professor, Department of Civil and Environmental Engineering, NUS*
Er. Chong Kee Sen, *Director, Engineers 9000 Pte Ltd*
Prof Chou Siaw Kiang, *Co-Lead for Energy Efficiency, Centre for Energy Research and Technology, NUS*
Mr Dalson Chung, *Director, National Environment Agency*
Mr Joe Eades, *Managing Director, Ispahan Group, Singapore*
Dr Goh Yang Miang, *Co-Chair, IES Health & Safety Engineering Technical Committee*
Dr Ho Kwong Meng, *Senior Consultant, Coastal Engineering, Surbana Jurong Consultants Pte Ltd*
Er. Ho Siong Hin, *Divisional Director, Occupational Safety and Health Division, Ministry of Manpower*
Mr George Huang, *Past President, Singapore Manufacturing Federation*
Er. Edwin Khew, *President, The Institution of Engineers, Singapore*
Mr Loy Chee Hiang, *Director, Seasky Netjoy Pte Ltd*
Dr Moh Chong Tau, *Former President and CEO of Makino Asia Pte Ltd*
Er. Timmy Mok, *Sr. Principal, T.Y. Lin International, Singapore*
Mr Ong Geok Soo, *Fellow, The Institution of Engineers, Singapore*
Mr Queek Jiayu, *Secretariat, The Institution of Engineers, Singapore*
Er. Seow Kang Seng, *Council Member, The Institution of Engineers, Singapore*
Er. Seow Tiang Keng, *Fellow, The Institution of Engineers, Singapore*
Mr Mervyn Sirisena, *Chairman, Aerospace Engineering Technical Committee*
Mr David So, *Vice President (Engineering), SIA Engineering*
Er. Tan Ee Ping, *Honorary Fellow, The Institution of Engineers, Singapore*
Er. Tan Seng Chuan, *Regional Managing Director, Ramboll Group*
Mr Desmond Teo, *Secretariat, The Institution of Engineers, Singapore*
Dr Sanjay Thakur, *Quality and Business Improvement Manager, Rolls-Royce*
Ms Wan Siew Ping, *Lead, Engineering Design, Simulation and Proto-typing Initiative, Singapore Institute of Manufacturing Technology*

Mr Philip Yeo, *Chairman, SPRING Singapore*
Dr Zhou Yi, *Assistant Professor, Singapore Institute of Technology*

Agency for Science, Technology and Research
Building and Construction Authority
Defence Science and Technology Agency
Economic Development Board
Energy Market Authority
Housing and Development Board
Info-communications Media Development Authority
JTC Corporation
Keppel Offshore and Marine
Land Transport Authority
Maritime and Port Authority of Singapore
PUB
SMRT
Singtel
ST Aerospace
ST Electronics

Photo Credits

Disclaimer

Every reasonable effort and care has been taken to trace the ownership and to any copyrighted material used in this book. It is not IES' intention to infringe on anyone's copyright for the photos, images or diagrams used. The publisher welcomes any information that clarifies the copyright ownership of any unattributed material displayed.

We would like to take this opportunity to thank the following organisations/persons for granting IES and the publisher to reproduce the photos/images/diagrams found within this book:

Building and Construction Authority
Changi Airport Group
Christiani & Nielsen (Thai) PCL
Economic Development Board
Energy Market Authority
ExxonMobil Singapore
Hong Leong Building Materials Pte Ltd
Housing and Development Board
Keppel Offshore & Marine Ltd
Land Transportation Authority
Ministry of Manpower

National Environment Agency
Nanyang Technological University
NUS Singapore Research Nexus, Singapore Photobank – Kelman Chang
Public Utilities Board
Remember Singapore
Sembcorp Marine Ltd
SGBuses.com
Singapore LNG Corporation Pte Ltd
Sunseap Group
ST Electronics
Workplace Safety and Health Council (WSHC)
Ms Wan Siew Ping

Photos reproduced with permission under Creative Commons:

Chapter	Creator	Original Title	CC License
1	mailer_diablo	EZ-Link Cardreader (Bus)	BY-SA 3.0
		ERPBugis	BY-SA 3.0
	Kelman Chang	MRT Gantry	BY-SA 4.0
	9V-SKA	2 C830 in KCD	BY-SA 3.0
	Littlearea	C830Interior	BY-SA 3.0
	Jnzl	Eco-link @ BKE	BY 2.0
	unkx80	Expressways and semi-expressways of Singapore 2	BY-SA 3.0
	epSos.de	Driving Cars in a Traffic Jam	BY 2.0
2	William Cho	Singapore River	BY-SA 2.0
	TimLee90	Marina Barrage Bridge	BY-SA 3.0
	AtelierDreiseitl	Singapore Bishan Park	BY-SA 3.0
5	Grps	The Winding Lanes of KB	BY-SA 3.0
	Jnzl	Dakota Crescent flats along the Geylang River	BY 2.0
	mailer_diablo	Bishan-SG	BY-SA 3.0
	Someformofhuman	Pinnacle @ Duxton, Singapore — 20100101	BY-SA 3.0
	Terence Ong	Supreme Court Building, Singapore	BY-SA 2.0
6	Sengkang	Singapore Changi Airport, Control Tower	BY-SA 3.0
	Pulkitsangal	Aerial view of Singapore Changi Airport and Changi Air Base — 20110523	BY 3.0

Contents

Foreword by DPM Teo Chee Hean	v
Foreword by Er. Edwin Khew	vii
Editor's Remark by Prof Cham Tao Soon	ix
Acknowledgements	xi
Chapter 1 Land Transportation	1
Chapter 2 Water	23
Chapter 3 Energy	39
Chapter 4 Manufacturing	55
Chapter 5 Buildings & Infrastructure	83
Chapter 6 Aerospace	103
Chapter 7 Infocomm Technology	125
Chapter 8 Offshore & Marine	135
Chapter 9 Health & Safety	151
Conclusion Engineers for the Future	169
Index	173

Chapter 1

Land Transportation

1. Introduction

For the past five decades, Singapore has experienced unprecedented development in its transportation landscape. What was once a small colonial settlement with few public transport options has spawned into a first-world transport system with extensive railway lines, comprehensive bus services to complement the railways and an intricate web of public roads.

This was possible given the government's huge emphasis on having an efficient transport system as a facilitator of economic growth and development, combined with the engineering ingenuity of her city planners.

Unlike many other major cities, Singapore does not have the luxury of having much land space and access to natural resources. Faced with such difficulties, city planners and engineers have had to explore innovative and ground-breaking ways to overcome these limitations.

1.1. *Brief history in time*

1970s

During the early years of independence, there only existed a small transportation unit within the Public Works Department (PWD) that oversaw the development of transportation in Singapore. Back then, public transportation was largely by bus and there were many bus services that worked independently from each other.

In a bid to improve the transportation landscape in Singapore to cope with her rapid population growth and economic needs, the government rolled out an enormous developmental project known as the State and City Planning (SCP) project. Work on SCP began in 1962 and took place over a span of 14 years, during which the master plan for land transport up to 1992 was developed.

1980s

One recommendation of the SCP was for railway systems to be constructed, with detailed studies to be conducted to assess when and how it should be deployed.

The Comprehensive Traffic Study, completed in 1981, reported that an all-bus system would be insufficient to meet our future transportation demands. Plans for a Mass Rapid Transit (MRT) system were then introduced, and major expressways such as the Pan-Island Expressway (PIE) and Bukit Timah Expressway (BKE) were built to improve road connectivity within the island.

1990s

In the early 1990s, different aspects of land transportation were concurrently overseen by four separate organisations:

- Registry of Vehicles managed all road vehicles in Singapore;
- MRT Corporation developed and managed the MRT system;
- Road and Transportation Division of the PWD developed roads and managed traffic, and;
- Land Transportation Policy Division of the Ministry of Communications oversaw all land transport-related policies.

Eventually, the Land Transport Authority (LTA) was formed in 1995 as the single agency responsible for all matters related to land transportation, including policy-making, land transport planning, land transport development and related regulatory functions.

2000s to present

New additions and extensions were made to the MRT system network, such as the construction of the North–East Line, Sengkang–Punggol Light Rapid Transit (LRT), Changi Airport Extension and Circle Line. In addition, there were comprehensive reviews on land transport governance and operational policies to further improve the efficiency and sustainability of our land transportation system.

2. Engineers' Contributions Towards Public Transport

Today, an approximate 7 million trips are made each day on the public transport system, with an average daily ridership of more than 4 million and 3 million for buses and rail transportation, respectively.

As the government tightly controls the percentage of private car ownership in land-scarce Singapore, the public transport network is designed to provide Singaporeans with a convenient, affordable, and efficient alternative. Throughout the years, engineers have contributed their expertise and effort to every part of the system, from the smallest components to the largest systems, from conception to fruition, toiling alongside their peers from various fields and domains to make life easier for all Singaporeans.

2.1. The MRT system

The first major revolution in Singapore's public transport system since independence was the introduction of MRT, a comprehensive network of railway systems that was conceptualised under the SCP project. Being a small island nation, there was a need to balance the use of land for roads vis-à-vis the use for housing and other developments.

Despite opposition from some members of the government due to the heavy investments involved (estimated to be around S$5 billion), the independently-conducted Comprehensive Transport Study revealed that putting more buses on the roads would lower traffic efficiency and result in more congestion. The MRT network was hence necessary to solve the nation's need for reliable public transportation as well as a sustainable solution to meet the increasing demands that accompany rapid population growth.

The construction of the then-largest public works project began in 1982 after three phases of study by the engineering and development team (Fig. 1). Priority was given to the North–South Line as it served central Singapore, which had a high demand for public transport.

On 7 November 1987, the first 6 km of the North–South Line (NSL) from Yio Chu Kang to Toa Payoh was officially opened for operation. The opening ceremony was inaugurated by then-Minister of Communications, the late Mr Ong Teng Cheong.

"This is like a 20-year affair from conception to delivery. Now the baby is born, to say that I am happy and pleased is an understatement."

Mr Ong Teng Cheong

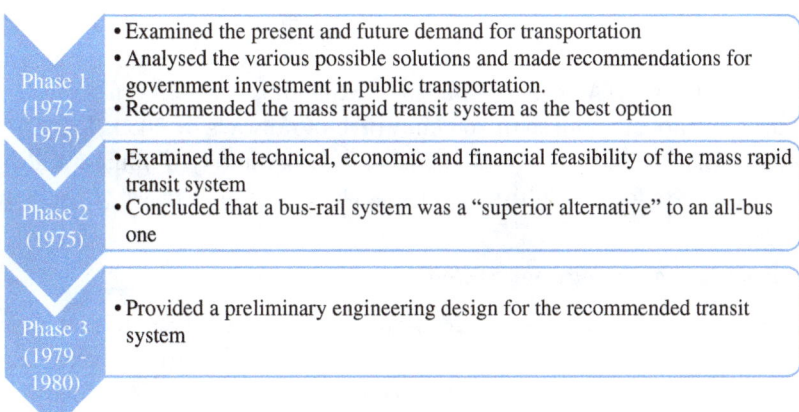

Fig. 1. Timeline of the three phases of development of MRT.

Fig. 2. Mr Ong inaugurating the MRT system in the opening ceremony.

This was followed by the stations from Novena to Outram Park, opened by then-Deputy Prime Minister Goh Chok Tong.

With the opening of a further six stations from Tiong Bahru to Clementi on the East–West Line (EWL) with City Hall and Raffles Place becoming interchange stations between the two lines, the system was officially launched on 12 March 1988 by then-Prime Minister, the late Mr Lee Kuan Yew.

2.1.1. *Expanding the rail network*

The NSL and EWL remained the backbone of Singapore's rail network for some 15 years until 2003, when the North–East Line commenced operations. In addition to these three lines, LTA is in the midst of expanding the MRT network further, with plans to establish more than 360 km worth of railways by 2030. The expansion will increase the projected daily ridership to approximately 6 million and help to strengthen the case for public transport as a choice mode of travel (Figs. 3 and 4).

2.1.2. *Integrated communication system*

An essential component to the smooth operation of the MRT system is its Integrated Communication System (ICS). ICS provides the means for carrying voice and data for smooth transit operations daily. This complex electronic engineering project is one of the important parts of a railway transportation system and is often updated to include the latest innovations in railway communications. In the case of Singapore, ST Electronics undertook the milestone project of our MRT ICS from conceptual design to commissioning.

Name	Commencement	Latest extension	Terminal	Number of stations	Length (km)
Operational					
North–South Line	7 November 1987	2014	Jurong East/ Marina South Pier	26	45
East–West Line	12 December 1987	2016	Pasir Ris Changi Airport /Tuas Link Tanah Merah	35	57.2
North–East Line	20 June 2003	2030	HarbourFront /Punggol	16	20
Circle Line	28 May 2009	2025	Dhoby Ghaut Marina Bay /Harbourfront Stadium	30	35.7
Downtown Line	22 December 2013 (Stage 1)	2025	Bugis/ Chinatown	6	4.3
Under construction					
	2016 (Stage 2) 2021 (Phase 3)	2025	Bukit Panjang /Expo	34	42
Thomson Line	2019 (Phase 1) 2020 (Phase 2) 2021 (Phase 3)	N/A	Woodlands North/Gardens by the Bay	22	30
Under planning					
Eastern Region Line	2021–2022	N/A	N/A	12	21
Jurong Region Line	2025	N/A	N/A	N/A	20
Cross Island Line	2030	N/A	N/A	N/A	50

Fig. 3. Additional lines under MRT expansion.

A brief overview of the ICS follows:

1. *Transmission System*

An optical fibre transmission system, equipped with fail-safes, serves as the backbone for all major communications within the MRT network. A system is connected to this backbone for audio and data transmission.

2. *Electronic Private Automatic Exchange (EPAX) System*

A 1000-line EPAX network is provided with subscriber telephones at all key locations in the station for administrative and maintenance personnel.

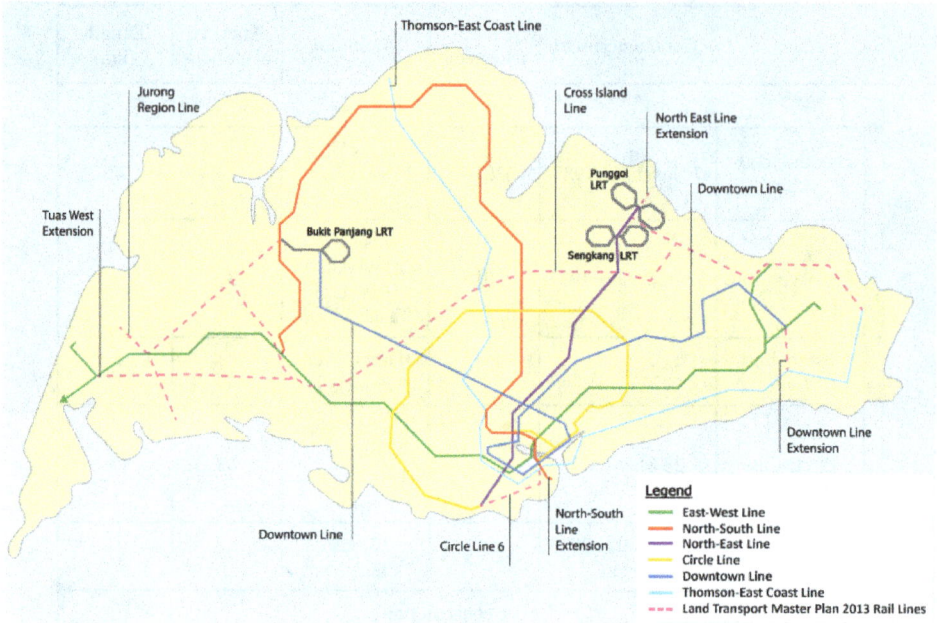

Fig. 4. MRT network in 2030.

3. *Direct Line Telephone System*

This system provides hotline facilities between the following telephones:

- Trackside Emergency Trip Station telephones to Optical Connection Controller (OCC)
- Station Controller to OCC and adjacent Station Controllers
- Maintenance Omnibus circuits linking up all the power substations and connecting Relay Rooms to point machines

4. *Closed Circuit Television (CCTV) System*

Fibre-optic cables and equipment are used to transmit video images from the stations to the controller. Colour CCTV cameras monitor key areas remotely from the OCC and locally at the station control room (SCR). A microwave link also transmits platform camera footage to the train driver cab so that the driver can ensure the safe entry and exit of passengers through the train doors.

5. *Radio Communication Systems*

The radio systems operate over the frequency range of 80–960 MHz. Elevated tracks and aboveground stations are covered by antennae on high masts. In the underground stations and tunnels, a total of 137 km of feeder cables provide radio coverage toward the trains. Communications with pagers and mobile telephones remain distortion-free even as the passengers transit between the underground and aboveground areas.

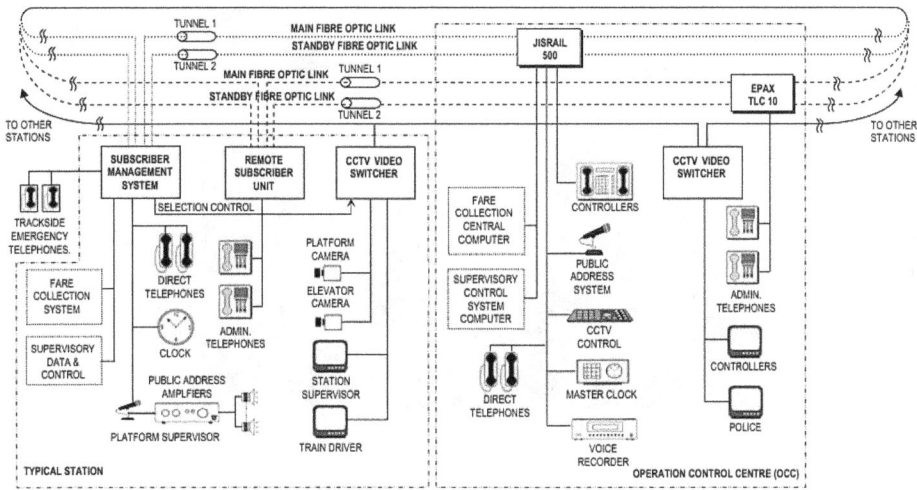

Fig. 5. ICS network diagram.

6. *Public Address (PA) System*

All stations are provided with PA coverage. The PA system at each station is divided into several zones such as concourse, platform and so on, and can be selected at will for address individually or collectively from the SCR or OCC.

Automatic noise monitor microphones are also provided to compensate for noisy environment through gain control.

7. *Other Systems*

Support facilities such as a system-wide synchronised master clock system and the voice recording of controllers' conversations are also provided.

Over the years, the communication systems, both on board the trains, as well as within the stations, have been improved continuously to ensure efficient communication at all levels, from the station to the staff and to the passengers (Fig. 5).

2.1.3. *Ticketing and fare system*

Another area of the public transport system that has been constantly improved by our engineers is the ticketing system.

The initial ticketing system in the 1960s involved commuters paying their fares through bus conductors, with bus tickets issued and hole-punched to indicate boarding locations.

With the advent of the MRT system, the Integrated Ticketing System (ITS) was introduced in 1990 to provide a common fare payment system on both rail and bus services. A magnetic ticket, termed the "fare card," contained a stored value which needed to be topped up at designated locations. The fare card was inserted into validating machines on buses and MRT stations and the corresponding selected fares were deducted.

Fig. 6. The present ez-link cards.

A significant milestone in the evolution of the ticketing and fare system came in 2002 with the Enhanced Integrated Fare System (EIFS). The system was developed in 3 years and replaced the magnetic fare cards with contactless smart cards, also known as EZ-link cards, which could be tapped against purpose-built readers on buses and at the MRT fare gates to allow smooth access and fare deduction.

The improvements to the ticketing and fare system have not only benefited commuters, but have also reduced the cost incurred by the government and the public transport operators. The improvements have resulted in the following:

- Faster commuter throughput for buses and trains as automatic fare calculation enables commuters to board buses and train stations more quickly, thereby reducing congestion and shortening the travelling time.
- Data warehouse for EIFS captures transaction details, which aids LTA in its strategic planning for public transport.
- Greater flexibility to set fares, for example, 1 cent increment instead of the original increment of 5 cents.
- Extended fare structure allows commuters to transfer between transport networks run by different operators easily without additional fare gates between said networks.
- Reduced maintenance costs of public transport, as contactless smart cards are more reliable and durable, with a failure rate of 1 in 25,000 transactions as compared to the 1 in 5,000 transactions with the older fare cards.
- Significant reduction in fare leakage for bus operators which is estimated to be in the region of $35 million annually.

2.2. *Improving train reliability*

With engineers' efforts in applying scientific, social, economic and practical knowledge, new technologies were developed to improve train reliability.

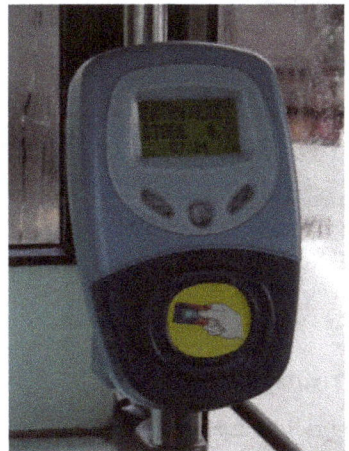

Fig. 7. Bus entry processor (left) and rail station entry processor (right) for the EIFS.

Fig. 8. Wheel Impact Load Detection (WILD) System.

In recent years, there has been an increase in the number of train disruptions which could have been caused by fatigue cracking due to prolonged usage and aging train tracks and defective wheels, among other things.

By adopting preventive maintenance approaches using new technologies, problems can be detected and rectified earlier. Some of the technologies utilised include the following:

- Wheel Impact Load Detection (WILD) System (Fig. 8)
    - Detects wheel defects
    - Alerts SMRT's Operation Control Centre so that maintenance can be carried out
- Linear Variable Differential Transformer (LVDT) System
    - Monitors track condition
    - Checks for any anomalies on the power-supplying third rail

2.3. *Ensuring the sustainability of our public transport systems*

Apart from enabling smooth operations and improving reliability through enhanced maintenance regimes, the long-term sustainability of Singapore's public transport system is a perennial concern for engineers.

2.3.1. *Energy sustainability*

The first issue at hand for engineers to tackle revolves around the difficulty of managing energy resources. The ever-growing demand for fuel and energy has led to many environmental issues, and much effort is and will be required to enhance awareness of the importance of environmental protection. Faced with such challenges, LTA has employed two main strategies for effective resource utilisation:

- Analyse and meticulously mine technological innovations for energy conservation benefits
- Institute design efforts in maximising energy efficiency

MRT

These strategies have been adopted, in the case of the MRT system, as described below:

1. *Electrical systems*
    a. Inverters are installed to recover the excess regenerative energy from braking by channeling the excess energy back into the internal AC HV network.
    b. Computer simulations are performed to determine the optimum location of inverters.
    c. This allows recovery of up to 5% of the total energy and improves receptivity of the traction network, which consequently reduces the rate of wear and cost of maintenance on the mechanical brakes of the train.
2. *Lighting system*
    a. Energy-efficient lighting such as T5 fluorescent technology or light-emitting diode (LED) lighting is considered in areas where lighting is required.

b. Possibility of utilising natural lighting is often explored to reduce energy consumption for lighting.
c. Different lighting control levels for different operational needs to allow appropriate selection of lighting to minimise consumption.

3. *Air-conditioning system*

   a. Carbon dioxide sensors are provided to regulate outdoor air supply by automatically adjusting fresh air supply rates to the MRT stations' public areas while ensuring CO_2 level is below 1,000 ppm, thereby reducing energy consumption without compromising air quality.
   b. Variable speed drives are provided to chilled water pumps and cooling towers to reduce the energy consumption by up to 0.4% during part load operation.
   c. Temperature sensors are provided in some mechanically ventilated rooms to ensure fans run only when necessary.
   d. Recycled water is used in the stations' air-conditioning systems.

4. *Platform Screen Doors*

   a. Prevent heat from the trains and warm, humid tunnel air from entering the station to reduce the cooling load in the station.
   b. Act as barriers to intrusions of tunnels and isolate stations from heat, dust, and air blast generated by the train movement.
   c. Savings in capital cost with a smaller plant and station footprint are also realised.

5. *Rolling stock*

   a. Weight is proportional to energy consumption.
   b. Use of aluminum alloy in the MRT trains improves weldability, mechanical strength, and corrosion resistance, allowing car body weight to be reduced significantly.

Fig. 9. The Alstom Metropolis C830 driverless trains for the Circle Line (left) and its interior view (right).

c. Case in point: For Circle Line trains, the weight of the trailer car and motor car was trimmed by 3.6 and 4.6 per cent respectively, by reducing amount of glass, utilising skeletal cable trays, aluminium diffusers, and body side interior plates and lighter seats.
d. Efficiency of traction equipment is enhanced by using Insulated Gate Bipolar Transistor (IGBT) inverters and permanent magnet motors that have significant weight savings and enhanced performance.
e. Medium frequency transformer which is lighter and can operate at higher frequencies is used to prevent high losses at low frequencies.
f. A high acceleration rate is used followed by a longer coasting period and then a longer braking period at a high deceleration rate to save energy.
g. Regenerative braking is used to recover braking energy.

6. *Alignment of the track hump profile*

a. An upslope hump is used to slow trains down when entering the station to reduce braking energy and downslope hump is used to facilitate acceleration of the trains when departing the station.

7. *Control of escalators*

a. Inverter system is provided to conserve energy by reducing the operating speed of 0.75 m/s to the standby speed of 0.2 m/s when escalators are detected by sensors to be at a no-load state.

Bus

With an average daily ridership of over 3 million rides and more than 300 bus services covering the entire island, buses are an integral part of the public transport network.

In line with the government's push for a cleaner and greener environment, innovative ideas to reduce the environmental impact of Singapore's buses have been studied. One such idea is GreenLite, a joint project between NTU and Beijing's Tsinghua University.

The eco-friendly bus does not run on fossil fuel. Instead, it is powered by a hydrogen fuel cell and lithium-ion battery combination, resulting in zero carbon emissions and only emitting clean water. Unlike conventional buses, GreenLite is not powered by a combustion engine, which makes travelling on it a pleasantly quiet ride (Fig. 10).

2.3.2. *Land usage sustainability*

Being a small country with many areas of development requiring land use, optimal land use will help to ensure that Singapore is able to meet the increase in public transport demand. One solution to our limited land resource is to build roads and the MRT network underground (Fig. 11).

Fig. 10.   GreenLite Bus — Singapore's first green bus.

Underground MRT Line — Circle Line

The year 1982 marked the beginning of Singapore's underground MRT network. Engineers played an important role in creating the technical specifications that made it viable and safe for both the workers who worked on the project and commuters.

Singapore welcomed its first fully underground orbital railway line, Circle line (CCL) in 2009. It is the nation's fourth MRT line since independence. This new addition further enhances connectivity by linking the inner suburban areas of the city together. Being built fully underground, it also allows land to be freed up for other uses.

The Circle Line runs through the busiest roads in Singapore and has interchange stations with the East–West, North–East and North–South Line. It connects the eastern and western areas directly to northern suburbs without having to pass through busy interchange stations such as City Hall and Raffles Place.

Construction of this additional line demanded extensive excavation of a tunnel stretching 33 km long. It was especially challenging, as the tunnel was positioned very close to buildings, pipelines, power cables, gas lines, and other underground stations.

Building Challenges

The construction of the Circle Line was a feat that required innovation in design, engineering, and construction. Some of the challenges faced during its

Land Use	Planned Land Supply (ha)	
	2010	2030
Housing	10,000 (14%)	13,000 (17%)
Industry and Commerce	9,700 (13%)	12,800 (17%)
Parks and Nature Reserves	5,700 (8%)	7,250 (9%)
Community, Institution and Recreation Facilities	5,400 (8%)	5,500 (7%)
Utilities (e.g. Power, water treatment plants)	1,850 (3%)	2,600 (3%)
Reservoirs	3,700 (5%)	3,700 (5%)
Land Transport Infrastructure	8,300 (12%)	9,700 (13%)
Ports and Airports	2,200 (3%)	4,400 (6%)
Defence Requirements	13,300 (19%)	14,800 (19%)
Others	10,000 (14%)	2,800 (4%)
Total	71,000 (100%)	76,600 (100%)

Fig. 11. The land usage as of 2010 and the proposed usage by 2030. *Source*: MND.

construction include:

(i) Challenge 1: Soft Soil Conditions
Engineers faced soft soil conditions at the Promenade and Nicoll Highway stations. They also had to be cautious when boring the tunnel under the Kallang Basin. Various construction methods, such as top–down construction, were used to combat these conditions.

(ii) Challenge 2: Interlaced Tunnels and Cables
Circle Line tunnels would sometimes be built close to the Kallang–Paya Lebar Expressway (KPE) tunnel, subterranean power lines and the Deep Tunnel Sewage System (DTSS) near the upper Paya Lebar/Airport Road junction. Due to this problem, engineers had to pay more attention to ensure safety to avoid train tunnel collapse.

(iii) Challenge 3: Sticks and Stones
During the construction of the Circle Line, extracting 500 reinforced concrete and steel piles left over from previous construction projects was one of the many challenges. Also, at Mountbatten station, rocks were found at a fairly

high level and required explosives to break them up before further action could be taken.

The ever-present challenge of land scarcity means that underground work will be expected to become more commonplace in the future. Ongoing and upcoming additions to the underground MRT network include the Thomson Line, Eastern Region Line, Jurong Region Line, Cross Island Line, North–South Line Extension, Tuas West Extension, Circle Line Stage 6, Downtown Line Extension, and North–East Line Extension.

Underground Road — Marina Coastal Expressway (MCE)

In terms of road infrastructure, the same concerns with land use apply. Engineers face challenges in balancing the allocation of land for roads and for other purposes. Similarly, going underground to maximise available built-up area has been practiced in Singapore. An underground road project that was completed recently is the 5-km long Marina Coastal Expressway (MCE), which features a 3.5 km underground tunnel. The tunnel's 420 m stretch under the seabed, 150 m away from the Marina Barrage, was one of the engineering challenges in the construction of the tunnel, because large amounts of water were let out from the barrage from time to time. The ingenuity of the engineering team enabled all the challenges to be surmounted, and the expressway was inaugurated on 29 December 2013 (Fig. 12).

3. Engineers' Contributions Toward Road Management

With a growing population and increased road usage, proper road management is essential to managing road congestion. Viable and unique engineering solutions were developed and implemented to meet Singapore's need for efficient and effective traffic management. The continual growth in road traffic increases the need for solutions that monitor and control traffic in order to increase road safety, improve traffic flow, and protect the environment. Several systems that were implemented by our engineers include the Expressway Monitoring and Advisory System (EMAS) and Electronic Road Pricing (ERP).

3.1. *Road expansion (supply control)*

Pan-Island Expressway (PIE)

The oldest and longest of Singapore's expressways, the Pan-Island Expressway, connects Tuas in the western part of the island to Singapore Changi Airport in the east. Considered to be the first expressway in Singapore, the construction of the 42.8-km long expressway began in 1964 and was completed in various phases, including an extension of the western end of the expressway in 1992 to connect it to Kranji Expressway. Presently, the expressway serves as the connection between major regions across the island, including Tuas, Jurong, Bukit Timah, Toa Payoh, Kallang, Eunos, Bedok, and Tampines.

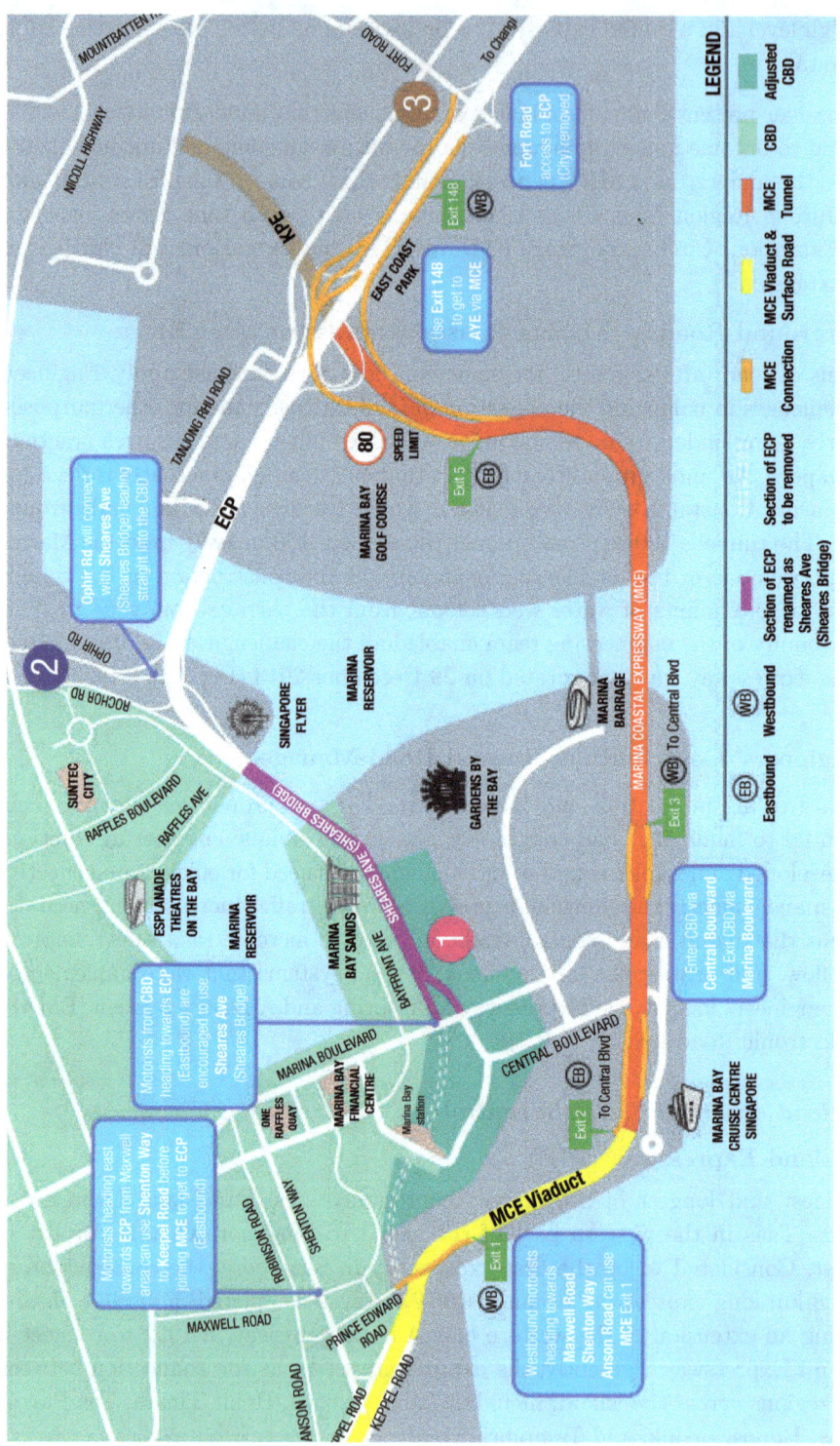

Fig. 12. Marina Coastal Expressway and Marina road network.

Bukit Timah Expressway (BKE)

Being one of the shortest expressways in Singapore at 11 km, BKE was initially constructed as the fastest way for motorists to travel between the north (Woodlands, Woodlands checkpoint, and Mandai) and PIE. During the 1990s, the east end of the Kranji Expressway (KJE) was connected to the BKE. During the 2000s, work was carried out to expand the section of the BKE prior to joining the Seletar Expressway (SLE). However, the BKE separates the Bukit Timah and the Central Catchment nature reserves.

For the past 20 years, this separation prevented the free exchange of biodiversity between the two nature reserves which upset the ecological balance of the two reserves. Recognising the importance of preserving these natural habitats for the future generations to enjoy, the government decided to construct a connection between the two nature reserves. The LTA, the National Parks Board (NParks), and other government and nongovernmental agencies were tasked to collaborate and construct the Eco-link@BKE.

Construction started on 30 July 2011 and was completed in late 2013 at a cost of S$16 million. Now, it serves as a reconnection between the two nature reserves and a visual treat for motorists travelling under it on the BKE (Fig. 13).

Central Expressway (CTE)

The Central Expressway (CTE) is the major expressway that links the north and the south of Singapore through the Central Business District (CBD). It connects the Seletar Expressway (SLE) at its junction with the Tampines Expressway (TPE) in

Fig. 13. Eco-Link@BKE.

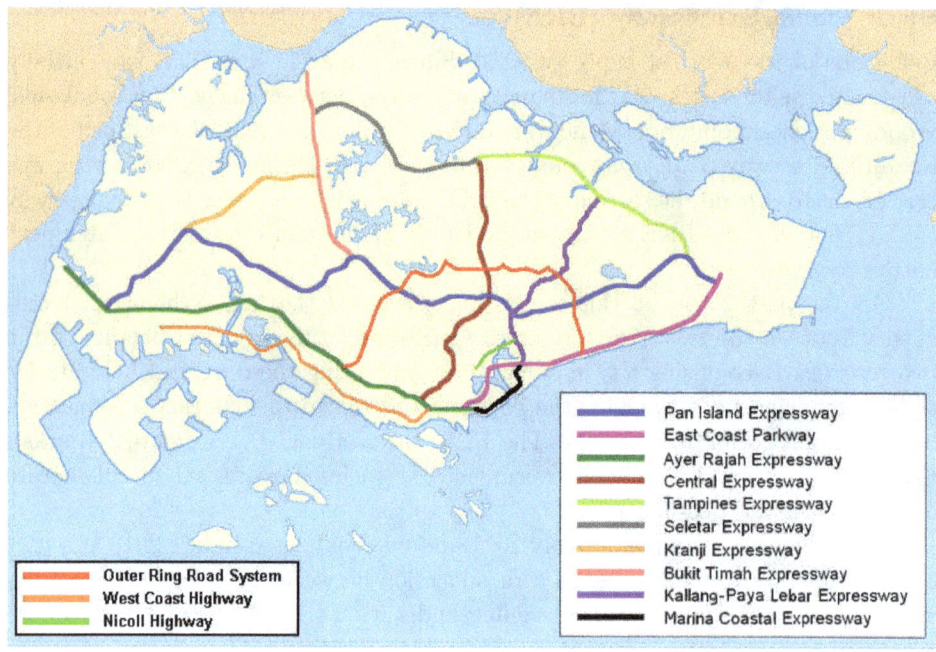

Fig. 14. Expressways and semi-expressways of Singapore.

the north, to the Ayer Rajah Expressway (AYE) in the south. Constructed in two phases, the CTE was opened in 1991. Sections of the road are laid underground, and these parts of the CTE formed the first underground highway of Singapore.

A new North–South Expressway (NSE) is being planned, running largely parallel to the central expressway (CTE) to cater to the increase in travel demand along the north–south corridor. This will help to alleviate the traffic load on the heavily utilised CTE and nearby major arterial roads such as Thomson Road and Marymount Road.

3.2. Better and more efficient transport management systems

As with other fast developing countries, Singapore faces challenges in terms rapid motorisation, urban planning, and ensuring that the public transport infrastructure keeps up with population growth.

There are three methods to resolve mobility problems, but each has its drawbacks:

1. Acquire land to expand road networks.

    - However, an increase in capacity is quickly saturated, leading back to the problem of congestion.

Fig. 15. Traffic congestion on an expressway in Singapore.

2. Transit-oriented development (TOD) adopted to reduce the dependence on cars for mobility by creating compact, pedestrian-friendly communities around public transport nodes.

   - Public transport cannot compete with cars that provide the comfort, privacy, and mobility.

3. Traffic management ensures traffic is well-distributed and smooth-flowing.

   - Peak hour traffic and traffic incidents can create turbulence to normal traffic flow.

In order to better combat these problems, engineers have developed Intelligent Transport Systems (ITS), which use information and communication technologies that integrate data from transport infrastructure, vehicles, and users to better manage transportation issues. ITS consists of three phases:

Phase 1: Enhancing Operational Efficiencies (Current)

- CCTVs are installed to provide remote surveillance of traffic intersections
- Intelligent sensors are added to detect traffic incidents, reducing resources required to look out for incidents
- Technology used to automate the process of vehicle identification, fee collection, and enforcement of congestion pricing
- Taxis equipped with GPS and mobile display terminals to track and communicate with them quickly.

Fig. 16. Electronic Road Pricing (ERP) gantry deployed to reduce congestion on busy roads.

Phase 2: Sharing Real-Time Information (Current)

- Providing fleet operators who manage public transport with real-time information will allow them to provide alternative modes of transport to motorists during congestion.
- Employing media and telecommunications companies to disseminate real-time information about traffic conditions via radio, SMS, web services, and TV.

    Allows motorists and travellers to travel via a combination of transport modes that best suit their criteria of shortest time, lowest cost, or convenience, to reach their destination seamlessly.

Phase 3: Interactive services (Current to near-future)

- Develop and promote more personalised information and interactive transport-related services.
    - ST Electronics engineers developed the Telematics Services Hub/oTTo-Go to provide an integrated platform to support interactive services and real-time information dissemination.
- Adopt open platform standards to develop value-added services and content targeted at drivers and other road users (e.g. navigation, traffic alerts, road routes, location-based services, multimodal transit services, infotainment services, etc.)

Another management solution implemented as part of the ITS was the Expressway Monitoring and Advisory System (EMAS) developed by ST Electronics for the LTA. The EMAS is a state-of-the-art electronic traffic management solution to help in the management of traffic incidents along the expressways. The operation of the EMAS provides motorists with timely traffic information and improves road safety and traffic fluidity through its detection, verification, and control functions.

4. Future Work

In view of the growing elderly population, Green Man Plus system has been installed at more than 500 pedestrian crossing to cater to the older generation. Train stations are equipped with at least one lift that is fully furnished with a tactile guidance system and wheelchair-accessible facilities.

The upcoming High-Speed Rail between Singapore and Johor Bahru will also be able to improve connectivity between Malaysia and Singapore.

5. Recommendation

(i) *Possible use of electrification*

A proposed recommendation to creating higher energy sustainability is through the use of hybrid vehicles such as hybrid electric vehicles (HEVs), plug-in HEVs (PHEVs), and elective vehicles (EVs). These vehicles are viable to replace gasoline internal combustion engine vehicles (ICEVs) to reduce total energy consumption and CO_2 emissions.

The following benefits can be achieved through the introduction of hybrid cars into the land transport system:

- HEVs can reduce up to 39% of energy consumption and CO_2 emissions, while PHEVs are able to reduce energy consumption by up to 46% and CO_2 emissions by up to 54%.
- A partial divorce between road transportation and the use of liquid fossil fuels.
- Decrease in Singapore's net emissions hence reduced the costs of processing these emissions.
- Growth of new technologies and new business sectors for economic growth.

References

Ho, S. (2014). *Mass Rapid Transit (MRT) system.* Retrieved from Singapore infopedia: http://eresources.nlb.gov.sg/infopedia/articles/SIP_2013-11-05_131443.html

Land Transport Authority. (25 April, 2017). *Building Challenges.* Retrieved from https://www.lta.gov.sg/content/ltaweb/en/public-transport/projects/circle-line/building-challenges.html

Nanyang Technological University. (20 July, 2010). *NTU gets GreenLite for Singapore's first truly eco-friendly bus.* Retrieved from http://news.ntu.edu.sg/pages/newsdetail.aspx?URL=http://news.ntu.edu.sg/news/Pages/NR2010_Jul20.aspx&Guid=444f54de-7c4e-432f-9f9c-714893613faf&Category=@NTU

ST Electronics. (2011). *Electrical & Mechanical Systems Contract for Circle Line and Circle Line Extension in Singapore.* Retrieved from http://www.stee.stengg.com/pdf/railway_systems/EM_CircleLine_Eng.pdf

ST Electronics. (n.d.). *Integrated Communications System for North–South, East–West MRT Lines, Singapore.* Retrieved from http://www.stee.stengg.com/pdf/railway_systems/ICS_sg-NSEW-Lines.pdf

Zhao, J., Shirlaw, J., & Krishnan, R. (2000). *Tunnels and Underground Structures: Proceedings Tunnels & Underground Structures, Singapore 2000.* Singapore: CRC Press.

Chapter 2

Water

1. Introduction

1.1. *History*

1960s

Singapore seceded from Malaysia in 1965. It soon became evident that water security would form a central element in the city-state's future evolution.

To begin with, Singapore has little land to collect and store rainwater; troubling issues relating to drought, floods, and water pollution were prominent.

During the early years of nation building, the banks of the Singapore River were a hub of trade and prosperity. Industrial enterprises sprouted up alongside the waterfront with heavy lighterage activity, the transfer of cargo between flat-bottomed barges of different sizes saturating river traffic. Boatyards, squatter settlements, street hawkers, and even farms proliferated. Many of Singapore's early settlers relied heavily on the Singapore River for employment and basic sustenance. Needless to say, the chronic pollution from these activities turned the river into an environmental hazard, detrimental to human health and destructive to biodiversity.

1980s

Mr Lee Kuan Yew, then Prime Minister, envisioned the need for a prosperous, lively riverside waterfront living along the Singapore River. In view of that, a major cleanup involving heavy dredging of the riverbed was employed to recover the debris piled up over decades of heavy use. Restoration of proper sewerage system, resettlement of squatters, relocation of heavy industries, and re-siting of street hawkers were carried out. Finally, the efforts of a decade-long project concluded in 1987, returning Singapore River its former charm and glory. Since then, many efforts have been made toward ensuring a sustainable water supply for generations to come.

Fig. 1. Singapore River today.

Present

At the present time, more attention is drawn toward water conservation. In this aspect, engineers have played a part in contributing to the national effort of ensuring a robust and sustainable water supply for Singapore.

1.2. *The four national taps*

Over the last 50 years, through strategic planning and investment in research and technology, Singapore has put in place a diversified and robust water supply through the four national taps. The water supply comprises (1) local catchment water, (2) imported water, (3) ultraclean, high-grade reclaimed water known as NEWater, and (4) desalinated water. This diversification has allowed the nation to close the water loop and helped her take a step toward water sustainability.

Local catchment water

Singapore, unlike other nations, incorporates the use of two separate systems for the collection of rainwater and used water. Through a comprehensive network of drains, canals, rivers, stormwater collection ponds, and reservoirs, this allowed urban stormwater to be harvested on a large scale, where it is treated, thus forming a sustainable source of water supply.

As of 2011, the water catchment area has been increased from half to two-thirds of Singapore's land surface with the completion of the Marina, Punggol, and Serangoon reservoirs. PUB, the national water agency, plans to increase Singapore's water catchment area to 90% in the long term.

Imported water

Singapore has been importing water from Johor, Malaysia, under two bilateral agreements. The first agreement expired in August 2011, and the second agreement will expire in 2061.

Ultraclean, high-grade reclaimed water known as NEWater

NEWater is high-grade reclaimed water produced from treated used water that is further purified using advanced membrane technologies and ultraviolet disinfection, making it ultraclean and safe to drink.

NEWater can now meet about 30% of the nation's current water needs. By 2060, NEWater is expected to meet up to 55% of Singapore's future water demand.

Desalinated water

Two desalination plants with a combined capacity of 100 million gallons per day can now meet up to 25% of Singapore's current water demand. With three new desalination plants in the pipeline, desalinated water will be able to meet up to 30% of Singapore's water demand in the long term.

2. Engineers' Contributions Toward Water Supply

Singapore has come a long way to her present state. In 50 years, we have turned our water vulnerability into a strategic asset. Despite that, we still inevitably face problems such as climate change which can affect our water supply.

Climate change is a global issue that may impact the sustainability of our water resources. In recent years, Singapore has been facing extreme weather conditions leading to problems such as flash floods and significant dry weather conditions along with bushfires and lack of water.

In the last few decades, our engineers have played an important role in developing and implementing innovative solutions not only to ensure a sustainable water supply, but also to mitigate and adapt to any challenges that might arise as we move forward.

2.1. *Ensuring sustainability*

In addition, PUB has also undertaken major engineering projects to help us ensure sustainability, such as the Marina Barrage and the Deep Tunnel Sewerage System (DTSS). In 2008, the Marina Channel was successfully dammed up to make the Marina Reservoir firm, creating Singapore's 15th reservoir and its first in the city centre. The barrage is also a tidal barrier to prevent flooding in the low-lying areas in the city. As the water is unaffected by high tides, its water level is kept constant, making it conducive for water sports and activities all year round.

On the other hand, the DTSS is a project which many of us may not take much notice of because much of it is built deep beneath the ground, as the name implies. These deep tunnels go even deeper than most of the MRT lines. When the DTSS is completed, used water will be conveyed from the households and industries to a centralised water reclamation plant for treatment before being further purified to become NEWater. Both the Marina Barrage and DTSS are engineering projects which had challenged our engineers to think out of the box.

2.1.1. *Water-saving devices*

Since 1983, the government has made it a mandatory requirement to install water-saving devices and fixtures in all non-domestic premises and common amenities of condominiums. Some of the devices include the constant flow regulators and self-closing delayed action taps.

The installation of low-capacity flushing cistern is also made mandatory since 1997, and in 2009, it is mandatory to install only dual flush low-capacity flushing cisterns for all new premises undergoing renovation.

Fig. 2. Bottles of NEWater.

2.1.2. *NEWater*

NEWater is the pillar of Singapore's water sustainability.

NEWater is ultraclean, high-grade reclaimed water made by further purifying treated used water through a rigorous three-stage process. Going through microfiltration, the unwanted substances such as suspended solids, minute particles, disease-causing bacteria, and viruses are filtered out by membranes, leaving only dissolved salts and organic molecules.

At the reverse osmosis stage, the water passes through tiny pores in a semipermeable membrane which only allows very small molecules like water molecules to go through. This ensures that undesirable contaminants such as bacteria, viruses, heavy metals, disinfection by-products, etc., are eradicated. The end result is high-grade water that is free from viruses and bacteria and containing very low levels of salts and organic matters.

As an added safety measure, the high-grade water then undergoes ultraviolet disinfection, guaranteeing the purity of the product water. Some alkaline chemicals are then added to restore the pH balance of the water. The end product water is ultraclean and safe for drinking. Figure 3 illustrates the entire process.

2.1.3. *Marina Barrage*

Marina Barrage, a multipurpose barrage and reservoir, was the vision of Singapore's first Prime Minister Lee Kuan Yew, who, in 1987, had expressed the idea of damming up the Marina Channel to create a freshwater lake.

Using a system comprising of gates and pumps, Koh Brothers engineered the construction of the barrage using nine steel crest gates (Fig. 5) spanning across the 350 m-wide Marina Channel to keep out the seawater. Under normal conditions,

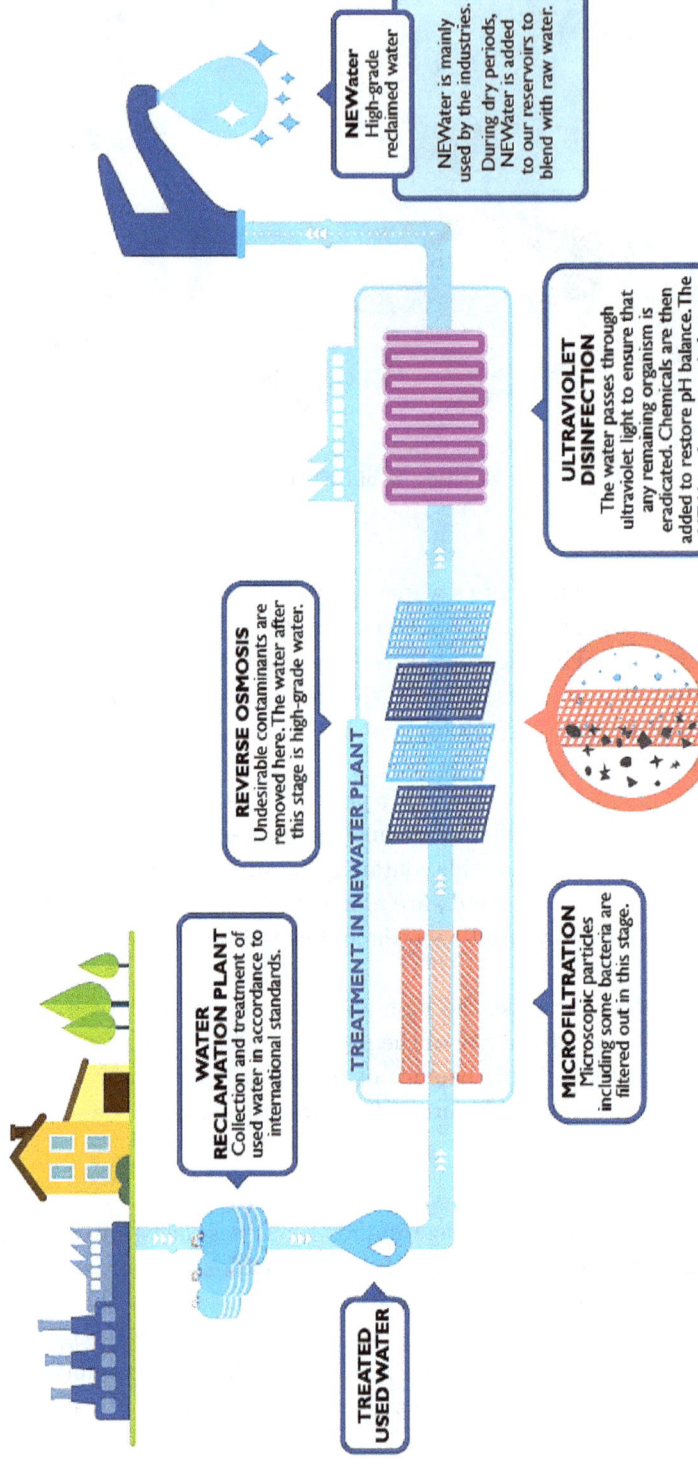

Fig. 3. NEWater Treatment Process.

Fig. 4. Marina Barrage.

Fig. 5. Marina Barrage Crest Gates.

the hydraulically operated gates will be closed. In the event of a heavy downpour coupled with a low tide, the gates will open to release the excess water into the sea. When heavy rain coincides with high tide, the gates will remain closed while the pumps will be activated to drive the excess water out to the sea.

In fact, the construction of the Marina Barrage has been considered as an engineering marvel due to its magnitude and complexity. To provide a dry working area, an 11 m-high cofferdam, which could withstand strong hydrostatic and wave pressures, had to be built. The presence of two layers of marine clay on-site added to the challenge. To ensure safety in such a difficult working environment, adequate instrumentation is also installed and monitored weekly. In addition, engineers also have to frequently monitor and deal with problems of water leakages and underground seepage.

Marina Barrage is dubbed as a three-in-one project. It is intended to boost our water supply, alleviate flooding, and become a lifestyle attraction.

Water supply

Built across the mouth of the Marina Channel, Marina Barrage is Singapore's 15th reservoir and the first in the heart of the city. With a catchment area of 10,000 hectares, the Marina catchment is the island's largest and most urbanised catchment.

Flood control

The Barrage is part of a comprehensive flood control scheme targeted to alleviate flooding in low-lying areas in the city such as Chinatown, Boat Quay, Jalan Besar and Geylang.

Lifestyle attraction

Being unaffected by tides, the Marina Basin is ideal for all kinds of recreational activities such as boating, windsurfing, kayaking, dragon boating, and so on.

2.1.4. Deep tunnel sewerage system (DTSS)

DTSS is a cost-effective and sustainable solution to meet the long-term needs of Singapore's used water collection, treatment, reclamation, and disposal. The DTSS uses deep tunnel sewers to convey used water entirely by gravity to centralised WRPs located at the coastal areas. This eliminates the use of intermediate pumping stations located across the island. At the WRP, the used water is then treated and further purified into NEWater, with excess treated effluent discharged to the sea through an outfall. The first phase of DTSS, covering the eastern part of Singapore, was completed in 2008 and at the time considered one of Asia's largest engineering projects. In the long run, it is expected to ensure NEWater sustainability, ensure efficient use of land, and grow Singapore's industry capabilities.

Although this is not the only underground work in Singapore, engineers face problems because underground construction is carried out in a complicated environment. The major tunneling challenges for this project include high groundwater table, very abrasive and mixed-face ground conditions, highly variable rock quality, and major expressways directly above the tunnels. The engineers managed to overcome these challenges and completed DTSS Phase 1 on schedule in 2008.

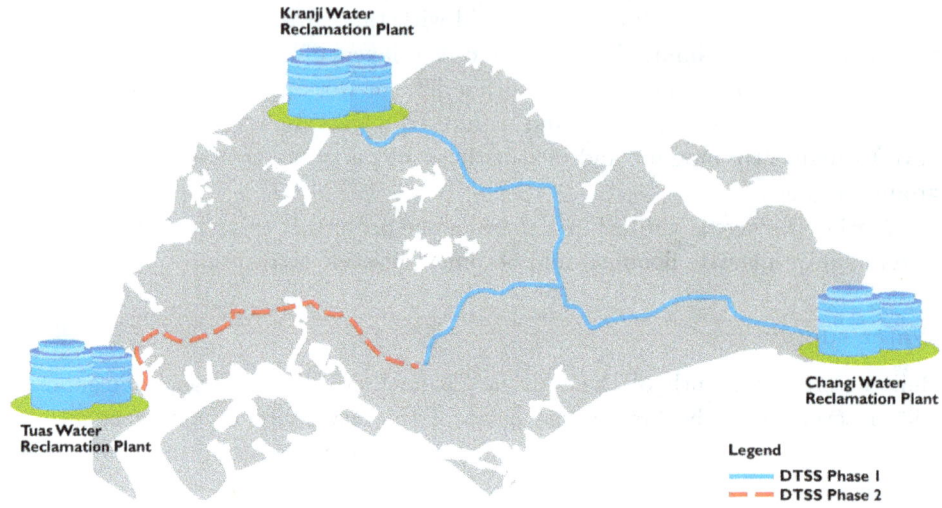

Fig. 6. DTSS Network.

Ensuring NEWater sustainability

DTSS is an important component of Singapore's water management strategy as it collects and treats every drop of used water and further purifies into NEWater. The largest NEWater plant in Singapore, to date, is built on the rooftop of the Changi Water Reclamation Plant, which is a unique feature of the plant to reduce footprint.

The integration of the NEWater plant with the DTSS facilitates large-scale water recycling and ensures the sustainability of NEWater for many generations to come.

Opened in May 2010, this NEWater plant at Changi has a capacity of 50 mgd. With this addition, together with the expansion of the existing three NEWater plants, NEWater now meets 30% of Singapore's total water demand.

Compact design and efficient land use

DTSS adopts a compact design and optimises land occupied by used water infrastructures. With the phasing out of existing intermediate used water pumping stations and the conventional plants, the previously occupied lands are now available for other developments. Under DTSS Phase 1, the compact design of the Changi Water Reclamation Plant requires only one-third the land area compared to a conventional plant. Also, there is no need for an odour buffer zone, as the plant modules are fully covered.

As such, DTSS results in a 50% reduction in land taken up by used water infrastructure once it is fully completed. During the 1990s, the land used to site the WRPs and the accompanying pumping stations was 300 ha. With DTSS Phase 1,

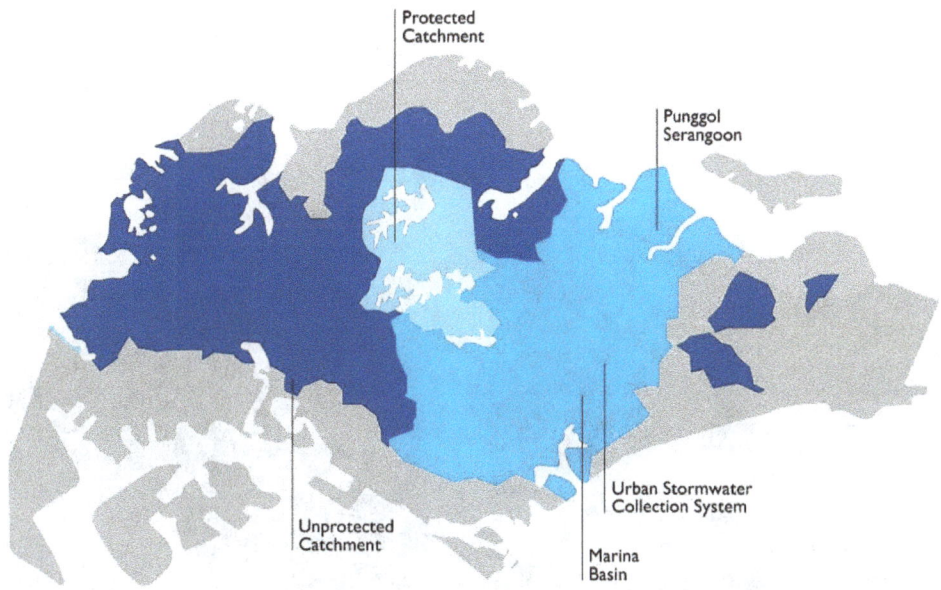

Fig. 7. Water Catchment in Singapore.

this area has shrunk to 190 ha. With the completion of DTSS Phase 2 in the future, the area will be further reduced to 150 ha.

Environmental protection

DTSS not only frees up land for higher value development, but it also enhances the reliability of the used water system as it minimises the risk of environmental contamination. This is important, especially since two-thirds of Singapore's land area is water catchment today (Fig. 7).

2.2. *Enhancing the quality of our living environment*

To enhance our living environment, efforts have been made to transform our waterways and reservoirs beyond their utilitarian functions into focal points for recreational and community activities under PUB's Active, Beautiful, Clean Waters (ABC Waters) Program. By bringing people closer to water, the program also creates opportunities for them to enjoy and bond with water, so that they can better appreciate and cherish this precious resource.

2.2.1. *Active, Beautiful, Clean waters (ABC Waters) Programme*

ABC Waters radically transforms Singapore's network of utilitarian drains, canals, and reservoirs beyond their traditional functions of drainage and water storage into beautiful and clean streams, rivers, and lakes.

Fig. 8. Activities to educate the public on the responsible use of our waterways and to appreciate nature were held at the revitalised Kallang River@Bishan-Ang Mo Kio Park project.

Such changes are made possible with the incorporation of engineering, science, landscape design, the behavioural framework of urban design, and a commitment to community involvement. These alterations offer recreation spots that are accessible and free enhancing the quality of life in an otherwise urbanised and fast-paced society.

ABC Waters involve much more than beautifying waterscapes and introducing activities. ABC Waters design features such as rain gardens, bioretention swales, and wetlands are not only aesthetically pleasing but also contribute to sustainable stormwater management. When implemented catchment-wide, these plants and soil medium with natural cleansing properties detain and treat runoff at source, hence helping to improve water quality that flows into the drains, canals, and, eventually, our reservoirs.

All in all, this seamless blue-green network is well integrated with adjacent land developments to continuously provide new community spaces and encourage new lifestyle activities to flourish in and around the waters.

One distinct example would be the **Kallang River at Bishan-Ang Mo Kio Park**. The project is the first of its kind in Singapore, where the traditional concrete canal flowing through the park is transformed into a naturalised river.

Under the ABC Waters Programme, bioengineering techniques comprising a combination of vegetation, natural materials such as rocks, and civil engineering techniques were used to stabilise the river banks to prevent soil erosion and form natural habitats that enrich the biodiversity of the park. This is the first time that such techniques are explored in a tropical climate. A test on 10 different types of soil bioengineering techniques was carried out prior to the commencement of the project, following which the most appropriate techniques were then applied at Kallang River@Bishan-Ang Mo Kio Park.

With the river naturalised, wildlife has flourished in the area. There has been an increase in the species of birds, butterflies, dragonflies, and damselflies sighted. The latest addition is the otters, which contribute to biodiversity at the waterway.

The river channel was designed to fulfil various functions. During heavy rain, the park land that is next to the river doubles up as a conveyance channel, carrying the flow downstream. On dry days, the flood plain is used as a recreational space, enabling multiple land uses within the park and creating more spaces for the community.

Driven by the vision to enhance the liveability of our living environment, Singapore will continue to undertake the challenge to transform itself into a City of Gardens and Water.

3. Future Development

3.1. *Tapping on R&D and new technology*

To keep Singapore's water supply sustainable, engineers tap on R&D and technology to try and find new and innovative solutions.

3.1.1. *Fish activity monitoring system*

The fish activity monitoring system is a successful home-grown technological solution. This technology serves as a simple early warning system where abnormal behavior of the fish indicates changes in water quality, prompting the system to send an alert to a central monitoring location. This technology acts as a front line of defence against water contamination, enabling a faster response to changes in water quality.

3.1.2. *Smart Water Grid*

PUB continually explores efforts in water innovation that will help minimise the losses of water due to leaks and ensure good water supply 24/7 to customers.

PUB, in partnership with MIT-Center for Environmental Sensing and Modeling (CENSAM) and Visenti Pte Ltd, has developed the Smart Water Grid, a network of wireless sensors installed in water supply mains across Singapore, which functions as a real-time platform to monitor water pressure, flow, and quality.

A total of 320 wireless sensors has been deployed across 5,490 km of potable water pipelines that reach 1.4 million customers. Each sensor station has a telemetry unit that transmits the sensor data wirelessly back to a central system for processing

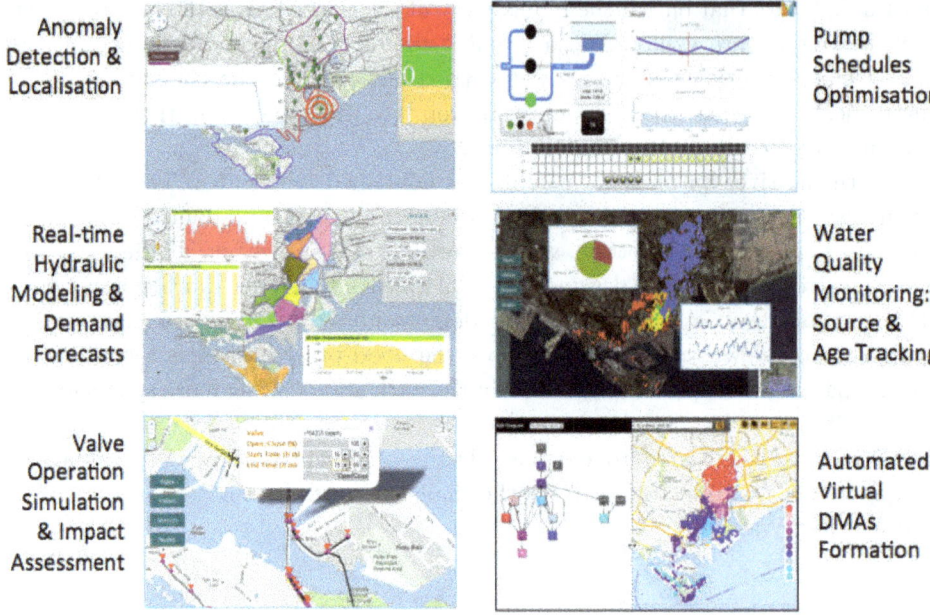

Fig. 9. Capabilities of the Smart Water Grid.

and analysis. In the event of any irregular readings and analysis, the system will notify PUB to carry out the necessary repair or recovery work.

The Smart Water Grid system also provides tools such as network simulation that help in decision-making processes, for example, analysing the potential impact of an operational event on the water supply network, therefore enhancing PUB's operations and the efficiency of water supply to consumers. The system also allows early detection of network occurrences with the objective of reducing response time, as well as minimising the impact and inconvenience to consumers.

3.1.3. *Deep Tunnel Sewerage System Phase 2*

A highlight of DTSS Phase 2 will be the new Tuas Water Reclamation Plant (WRP) which will incorporate technologies to improve its energy efficiency and manpower requirements. Besides its compact design, the new Tuas WRP will also be greener than the existing WRPs, producing more biogas for power. It will generate less sludge and therefore reduce the cost of sludge disposal. In the long run, DTSS Phase 2 will enhance water sustainability by helping to raise the water recycling rate from 30% to up to 55% of our total water demand.

3.1.4. *Keeping cool while saving water*

In hot and humid Singapore, air-conditioning systems are used to keep commercial buildings, offices, hotels, and hospitals pleasantly cool, and these systems rely on cooling towers to reject heat from the building by evaporating water. As a result, copious amount of water is lost daily through these cooling towers.

One of the current research projects funded by PUB is the recovery of waste heat in air-conditioning systems. The ultimate aim of the project is to recover the low-grade waste heat from the air-conditioning system, which otherwise is lost to the environment due to evaporation from the cooling tower, and put it to good use. Rather than rejecting all the heat back into the atmosphere, the project seeks to recover the waste heat to generate power or to meet the building's hot water needs. Consequently, the cooling tower will consume less water thanks to the reduced heat load to be discharged.

Singapore-based company Natflow is the lead project collaborator and has more than a decade's experience in heat recovery systems. By supporting the project, PUB hopes to advocate the use of waste heat recovery systems to accrue water savings and achieve net zero water wastage while still helping to keep the buildings cool.

Water engineering is vital to our everyday lives. Engineers not only deal with flooding, but also with water distribution and sewerage functions. Engineers find ways to conserve water and boost our water supply in order to meet the demand now and in the future. In Singapore, engineering projects are challenging because of our limited land, thus moving all construction of water projects underground in a complex environment. Climate change is a serious problem globally, and an engineer now takes on a new challenge of ensuring the security of our water supply by dealing with weather variability and uncertainty.

Singapore has come a long way in terms of water management. This improvement over the decades would not be possible without the hard work put in by the engineers of various fields. Engineers contribute at a national level in every major projects, building our nation.

References

Public Utilities Board. (2016). *Our Water, Our Future*. Singapore.
Schmid, S. (2012). *Catching Rainfall in Marina Bay: Water, Necessity, Policy and Innovation in Singapore*. Retrieved 9 March, 2015, from The Wharton School, The University of Pennsylvania | Initiative for Global Environmental Leadership: https://igel.wharton.upenn.edu/wp-content/uploads/2012/09/IGEL-Marina-Barrage.pdf
Singapore Economic Development Board. (2016). *Environment and Water*. Retrieved from EDB Singapore: https://www.edb.gov.sg/content/edb/en/industries/industries/environment-and-water.html
WSP. (2017). *Singapore Deep Tunnel Sewerage System*. Retrieved from http://www.wsp.com/en-GL/projects/singapore-deep-tunnel-sewerage-system

Chapter 3

Energy

1. Introduction — *Light of the South*

During World War II, Singapore was once known as the *Light of the South*. Though our nation has gained independence for 50 years, Singapore has yet to grow out of that given name. When night falls and the city lights up, streets around Orchard Road are filled with shoppers; parks are still filled with cyclists and runners; and the night view of Singapore never fails to make one slow down their footsteps to enjoy the scenic view of Singapore's skyline. The night life of Singapore is as busy as the day. This is made possible with the help of our unsung heroes to ensure a stable and reliable power supply to our nation.

Electrical and Energy (EE) engineers are part of the backbones to Singapore's economic developments. With the global concerns of climate change, many countries, including Singapore, are moving toward sustainable technology and design to reduce our contribution to carbon footprints. Hence, engineers from different fields work together to create and invent magnificent and sustainable products and services. Some of these products and services will be mentioned in the subsequent sections.

As population and automated industrial activities increase over time, energy demand is predicted to double by 2030. This further demands the creativity and skills of engineers to ensure supply-demand balance and maintain, or even upgrade, the current standards of energy supply in the future.

The contributions from our unsung heroes may soon be forgotten while many of us enjoy the fruits sown by them during the past decades. Hence, this chapter is a special dedication to our EE engineers.

1.1. *Singapore power supply history*

Besides the changing landscape of electricity generation, a full-scale rural electrification scheme was vigorously implemented in 1963. By 1974, electricity supply was available to most rural areas. There have been great improvement in the transmission and distribution networks together with the continuous improvements in the monitoring and control of these electricity supply networks over the decades.

1861–1862	The Singapore Gas Company was formed. At the same time, the Kallang Gasworks was built to supply piped gas from street lighting. Gas was produced from coal until 1958, when it was replaced with oil.
1905	Power station built in Mackenzie Road to supply electricity for trams.
1924–1927	St James Power Station, which was fired by coal, was built and began to supply electricity for Singapore's needs. It was decommissioned in the 1970s.
1963	Public Utilities Board (PUB) formed to supply gas, water and electricity to consumers.
1995	Singapore Power incorporated as a commercial entity, taking over the business of supplying gas and electricity from PUB.
1990s	Singapore began to diversify its energy sources, using natural gas to complement fuel oil in electricity generation. In 2002, oil accounted for about 51 per cent of electricity generated, natural gas for 44 per cent and waste incineration for the rest.
2001–2003	Electricity market was liberalised to let suppliers compete to provide power to about 10,000 non-residential consumers.
2006	The decision was made to import liquefied natural gas (LNG), which does not have to come from neighbouring countries unlike piped natural gas.
2008	Tuas Power announced it will build a steam-and-electricity plant that will run on biomass (plant matter) and coal for the region.
2009–2011	HDB announced a $31 million, five-year-trial of solar power at 30 precincts; using solar energy to power lights in common areas such as stairwells.
2011	Malaysian electricity group Tenaga Nasional approached Singapore regarding the purchase of electricity to tide it over in the event of shortages. Previous emergency shortages have seen both countries share electricity supply via two submarine cables linking Malaysia's grid with Singapore's at Senoko.

Fig. 1. A brief summary of the history power supply in Singapore.

2. Engineers' Contributions Toward Energy Sustainability

Singapore electricity supply was generated by burning fossil fuels, mainly fuel oil, since the start of the first power station in 1905. Piped Natural Gas was first introduced from Malaysia to replace the dirtier fuel oil. As more piped Natural Gas was being brought in from Indonesia, Singapore has diverted its power sources toward cleaner Natural Gas to reduce its carbon footprint (Fig. 2).

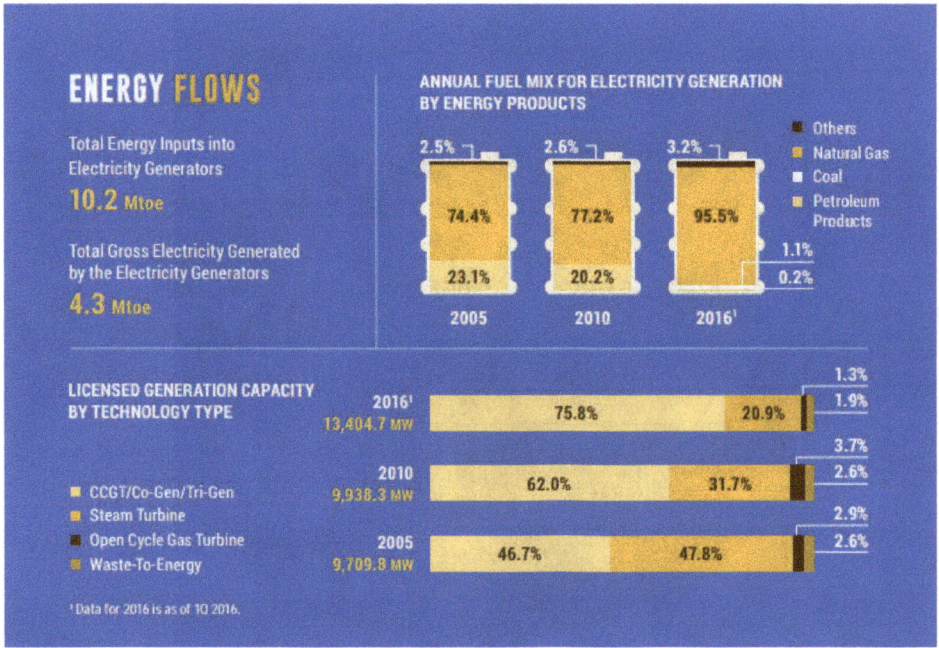

Fig. 2. Singapore Power Supply sources.

The motivation toward a sustainable living allows engineers to think and act out of the box. In the past few years, efforts have been made to introduce rooftop solar photovoltaic panels or solar energy on both industrial and public housing estates (Fig. 3) to further reduce usage of the dirtier fossil fuels. Singapore Marina Barrage Solar Park is one of the successful stories of using solar energy to power the lightings of its gallery. Housing & Development Board (HDB) has also welcomed the idea of green technology and improvised solar technology on the roof of existing buildings in Punggol. However, as Singapore has no indigenous energy resources such as wind energy resources and there are limited rooftop spaces available, she has to continue to depend on energy imports, such as liquefied natural gas (LNG), which is much cleaner than fuel oil, to support her overall energy demands.

2.1. *Improving power supply efficiency and reliability*

Singapore has one of the best electricity transmission and distribution networks in the world. The network performance of the national power grid compares favourably with the best internationally. To ensure that the network continues to meet the increasingly stringent needs of our consumers, continuing efforts and investments

Fig. 3. Solar panels atop HDB blocks in Jurong.

have been made to enhance the transmission and distribution of reliable and quality power.[1]

Good-quality underground cable system

A major factor contributing to good reliability and quality is the fully underground cable system that we have in place. Underground cables are installed to replace overhead lines during the initial two decades of our nation building in order to prevent damages caused by inclement weather and make the city look neater and tidier. However, the underground installation of cables presented another problem that affects the reliability of our power supply.

Since the 1990s, bigger and more powerful excavators were used for earthworks. These excavators were able to dig deeper and faster as compared to traditional manual labour and contributed to damages to the underground electricity cables, as shown in Fig. 4.

---

[1] http://www.singaporepower.com.sg/irj/go/km/docs/wpccontent/Sites/SP%20PowerGrid/Site%20Content/Resources/documents/Working%20Together%20to%20Prevent%20Cable%20Damage.pdf

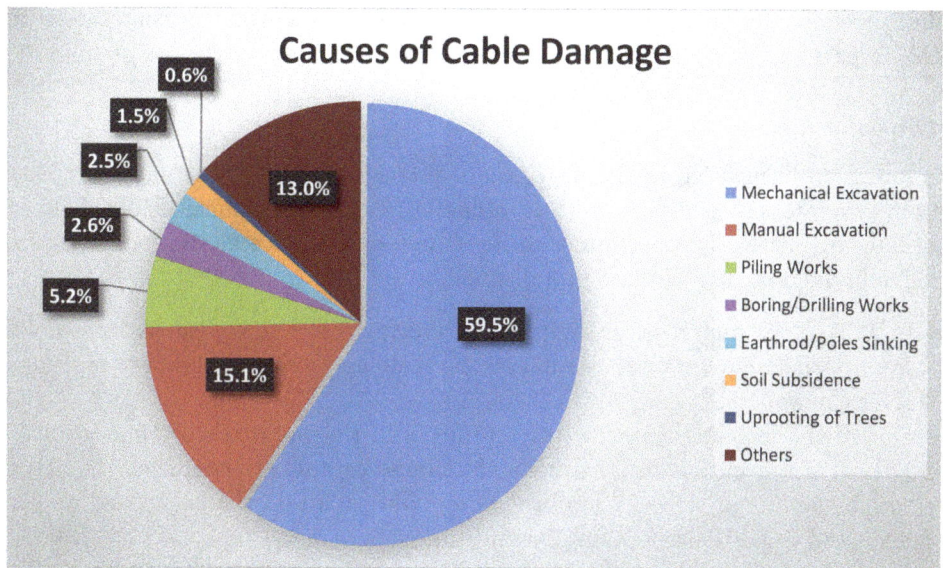

Fig. 4. Causes of cable damage in Singapore (*Source*: Singapore Power).

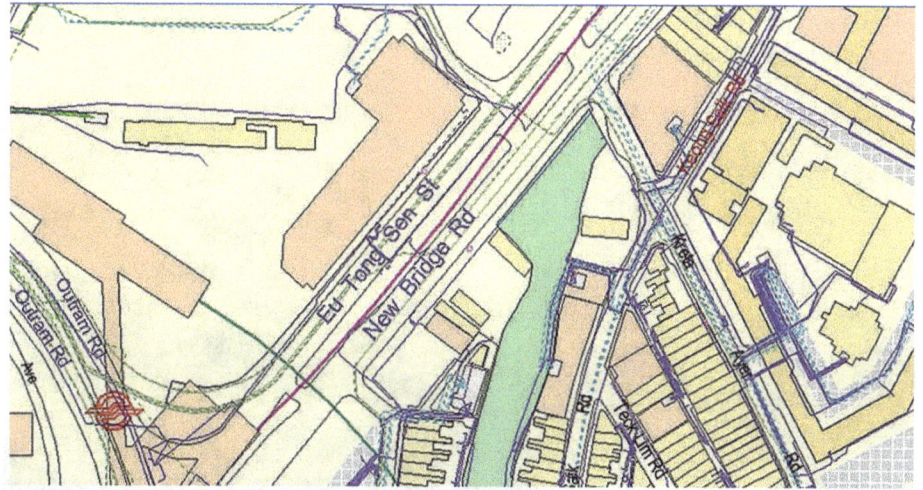

Fig. 5. Example of a cable plan indicating the locations of the underground cables.

Engineers of the then Electricity Department from the Public Utilities Board (PUB) came up with a solution to this issue by having a computerised mapping system (Fig. 5) of the underground networks of cables and pipelines. This led to more accurate access to the location of cables and pipelines, allowing construction work to be planned carefully beforehand to prevent damage to the power cables and pipelines. Engineers were also deployed to educate earthwork contractors on the preparatory works to prevent such damages.

Crossing the sea

Another problem faced in the provision of power supply is in carrying electricity from generators to end users such as refineries and petrochemical plants that are situated on offshore islands. Our engineers came out with several solutions to ensure reliability and quality of these power supplies.

- In 1969, PUB undertook a multimillion dollar project to lay submarine cables from Pasir Panjang Power Station on the mainland to the oil refinery on Pulau Ayer Chawan.
- One of the largest submarine cable projects in the world at the time involved 132 km of conventional submarine underground cables in the sea from the island to mainland, and some 27 km of underground cables on the island and on the mainland to the Pasir Panjang Power Station.
- In mid-1980s, a submarine cable tunnel was built at the offshore Pulau Seraya Power Station at about 20 m below sea level. With a terminal building at either end, the tunnel was designed to hold transmission cables as well as alternative potable water supply for the power station (Fig. 6).
- Building of the tunnel has saved millions of dollars as compared to laying conventional submarine cables.

Fig. 6. Internal view of the submarine cable tunnel linking Seraya Power Station.

2.2. Enhancing the nerve centre

Power System Control Centre (PSCC) of the Power System Operation Division (PSOD) of the Energy Market Authority is the nerve centre of the electricity generation and transmission system. It monitors the operation of the power system in Singapore and ensures the security of supply of electricity to consumers.

Until the 1960s, operation of the electricity network was carried out manually onsite, and this meant that problematic equipment in unmanned substations could go undetected until a serious fault developed. Additionally, a significant number of manpower was required to attend to the various substations that were located all over the island.

- In 1979, the PUB introduced computerised control with a new Energy Management System at the Ayer Rajah substation. Called SCADA, abbreviation for Supervisory Control and Data Acquisition System, it enabled remote control and monitoring of the transmission networks.
- PSOD operates the mission critical Energy Management System, which has been upgraded several times over the years to keep it up to date. PSOD also carries out power system studies, and assessment of the impact of new generating plants as well as the transmission expansion plans of electricity and gas transmission licensees.

3. Greener City, Greener Energy

As Singapore has no indigenous natural resources for energy supply, investment in renewable, and cleaner energy sources is one of the long-term solutions to ensure sustainable nation progress.

Currently, about 95% of Singapore's electricity is generated using Natural Gas imported from Malaysia and Indonesia. To enhance Singapore's energy security and to enable the import of gas from all over the world the liquefied natural gas (LNG) terminal on Jurong Island commenced operations in May 2013 (Energy 101, 2013).

The terminal has an initial capacity of 3.5 million tonnes per annum (Mtpa), and this capacity increased to 6 Mtpa in 2014 when a third tank and additional jetties and regasification facilities were added. The terminal's throughput capacity would further rise to 9 Mtpa when the fourth tank and its related regasification facilities are constructed.

3.1. Engineering challenges faced

EE engineers have faced many challenges in the past as mentioned in the previous sections. Some of the major challenges include:

- Rapid development of new infrastructures, such as newer and larger power stations, higher voltage larger capacity of transmission networks, rural electrification, etc., and introduction of new technologies, such as Energy Management System, SCADA, smart meters, etc. to cope with the rapid growth in demands and service

quality of electricity supply in the initial nation-building phase and subsequent rapid economic growth phase;
- Introduction of intermittent renewable energy sources, such as solar PV installations and less reliable embedded generation plants; and
- Introduction of wholesale electricity market that solely schedules generation capacity based on price bidding and leaves the long-term planning, real-time balancing, and dispatching of generation in the hands of engineers in the Power System Control Centre

In a special report mentioned in *The Straits Times* on November 8 2011, *Hooking up to Region's Powerhouse*, it highlights one of the potential issues engineers will need to solve if Singapore is able to import electricity from her neighbouring countries — the need to improve the technology used to transport electricity (Feng, Chua, & Cheam, 2011). However, besides the range of policy, regulatory and fiscal obstacles that have been well documented by ASEAN officials, main challenges faced by engineers are not the issue of energy lost during transmission from neighbouring countries, but a harmonised code or guidelines, appropriately trained engineers, legally empowered grouping of competent engineers to plan and coordinate cross-border transfer of energy that ensure reliable operation, performance, and safety in generation and transmission of electricity between countries for energy import and export via the proposed ASEAN Power Grid.

Other major challenges to be faced by EE engineers in the subsequent decades will include the following:

- To continue manage the critical assets in power generation, transmission and distribution of electricity in such a way to maintain, if not better, the current security and reliability of their operations and performance with more and more distributed generation and intermittent renewable energy sources, such as solar PV;
- To ensure that the power grid keeps in pace with the technological advancement being introduced in the smart nation initiative, particularly in deployment of electric cars, buses, etc., and smart sensors and meters; and
- To maintain the infrastructure to allow consumers freedom to choose their preferred electricity retailers without compromising their electricity services.

3.2. Energy fuels

3.2.1. Coal

Singapore's first utility plant to partially burn coal was opened on 27 February 2013. About 66% of its capacity will be fed by low-ash, low-sulphur coal, 16% by palm kernel shells and wood chips, and 18% by natural gas or diesel.

Future phases will add two more circulating fluidised bed boilers, a gas boiler, two steam turbines, and waste water treatment and desalination facilities.

When fully completed in 2017, the plant will be able to produce 160 megawatts of electricity and 900 tonnes of steam per hour.

Although some environmental groups have criticised it for burning coal which produces twice as much carbon dioxide per unit of energy as natural gas, the design of this power plant has ensured that the coal remains enclosed from barge to boiler in order to minimise coal dust. Filters to trap particulates and sulphur dioxide and coal burning technology to lower nitrogen oxide emissions incorporated in the plant design are able to meet the National Environment Agency standards, according to Mr. Lim Kong Puay, President and Chief Executive Officer of Tuas Power Ltd, the company that invested and operated this coal power plant.

The coal is stored in enclosed silos that hold 22,000 tonnes, or 2 weeks' supply. And energy-efficient processes, such as producing steam and electricity at the same time, mean most of the energy stored in the fuel is used rather than lost. Coal was chosen to diversify the plant's fuel mix for energy security and price stability. Coal prices have remained stable, while oil prices are volatile (Chua, 2013).

3.3. Liquefied Natural Gas (LNG)

The Singapore LNG terminal is a strategic and critical national infrastructure developed by the Singapore government. It allows Singapore to strengthen its energy

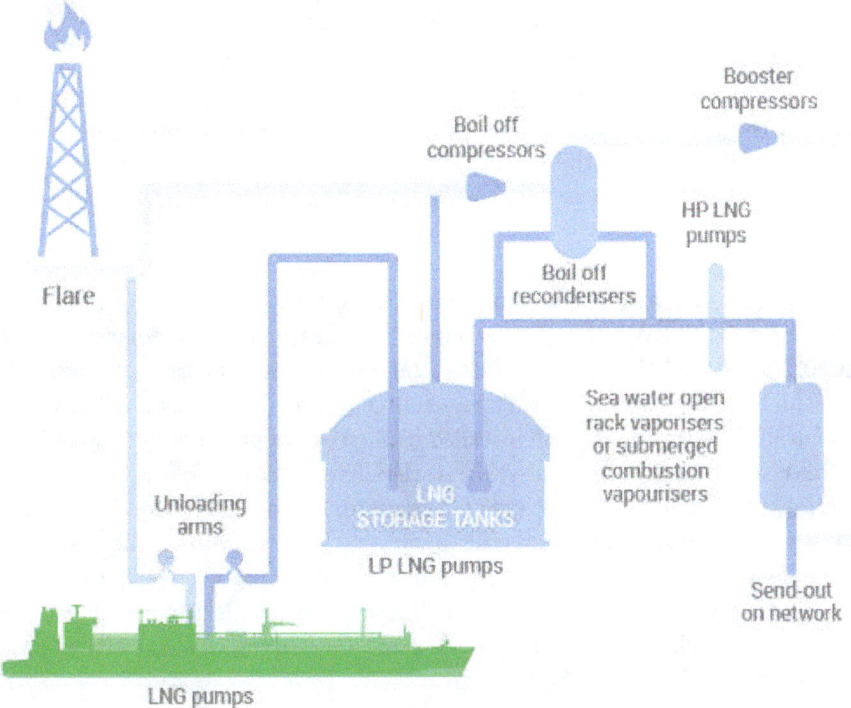

Fig. 7. Process chain of LNG system.

security by enabling LNG to be shipped to Singapore from all over the world and then re-gasified for use by the power generation companies and other industries. This reduces the nation's reliance on piped natural gas.

Beyond energy security, the terminal also serves as a platform and catalyst for LNG to potentially become the country's next major economic driver and for Singapore to establish itself as an LNG-trading hub for the region. It will help generate more LNG-related business opportunities and create new jobs in the energy sector.

The LNG terminal supports Singapore's carbon emission reduction efforts by enabling greater access to LNG, which is a less carbon-intensive and lower emissions fuel compared to oil or coal. Engineers implemented a number of unique/innovative and environmentally sustainable engineering approaches in this project. They include the following:

- The boil-off gas (BOG) compressors have a built-in third stage for sending high-pressure gas to pipeline. Most other terminals have a separate high-pressure BOG compressor.
- Variable frequency drives are used in some of the in-tank pumps, high-pressure LNG pumps and BOG compressors, to allow reduced energy usage. This is not common in other LNG terminals.
- The terminal has blast-resistant buildings to keep key facilities in the plant safe and under manageable control and asset protection.
- Gas safety case and safety management system were considered in the design, construction and operation to mitigate hazards unique to the Singapore LNG terminal (Fig. 8). The SLNG Gas Safety Case has won the prestigious Process Safety Award 2013 from the Institution of Chemical Engineers, UK.

3.3.1. Solar-power

Singapore is blessed with the geographical location where an average annual solar irradiation of 1,580 kWh/m^2/year is captured. This amount of solar radiation is about 50% more than that in temperate countries and thus makes solar photovoltaic (PV) technology a promising option as a source of renewable energy (EMA, 2010). As of end-2016, Singapore has an installed capacity of approximately 126 megawatt peak (MWp) for solar power, based on statistics from EMA.

Yet, due to the urban landscape of Singapore, space is restricted for the installation of solar panels. This causes EE engineers to work within the restrictions, creating efficient circuit systems to maximise the energy output and to deal with the intermittency of solar PV generation. Hence, solar panels are placed at the roof of Housing & Development Board (HDB) flats. One MWp of electricity is enough to provide for 200 four-room flats.

To maximise the potential of deploying solar energy when the technology becomes commercially viable, we need to invest in Research & Development and

Fig. 8. Aerial view of SLNG terminals.

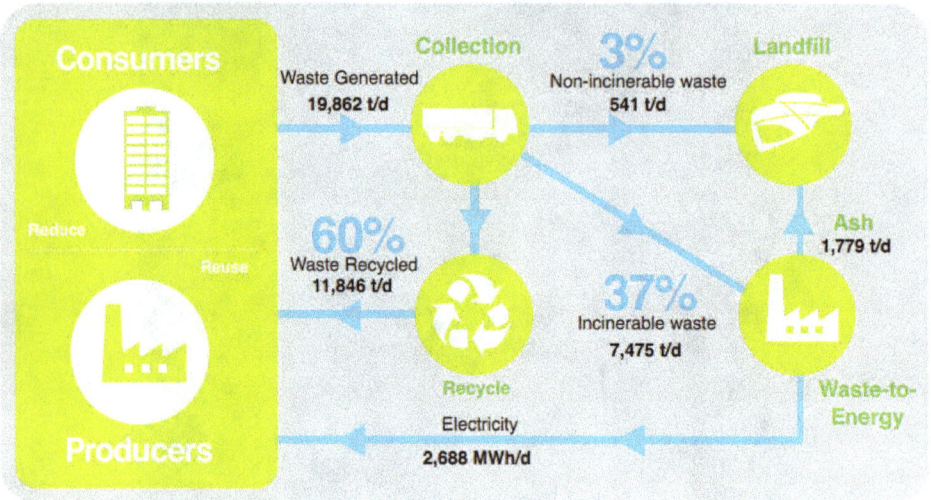

Note: Average daily figures for 2012

Fig. 9. Flow diagram of Singapore Waste Management.

develop more cost-effective ways to overcome the problem of intermittency associated with solar energy.

3.3.2. *Waste to energy — the use of biomass*

One of the biggest moves by Singapore to minimise the reliance on fossil-fuel generated electricity is through the waste-to-energy technique, where municipal wastes become a form of energy source. There are a total of four waste-to-energy incineration plants in Singapore, namely Tuas, Senoko, Tuas South and Semakau Landfill. As shown above (Fig. 9) the total amount of waste sent to incineration plants for energy generation is approximately 7,475 tonnes/day. The process of generating electricity from waste is shown in Fig. 10.

Besides having municipal waste as a form of energy source, animal waste is another potential energy supplier. In Singapore, biogases such as methane are obtained from animal wastes through anaerobic or fermentation process. Besides using animal waste, food waste can be a form of energy source. The treatment process of food waste to generate methane is similar to that used in animal waste.

4. Future Development

The creativity to find alternative energy sources as a form of power supply is endless. The international Energy Agency (IEA) estimates that the share of renewable energy in global electricity generation will increase from 20% today to almost a third in 2035. Singapore has a strong role to play here as the leading clean energy hub for the region, by partnering closely with business in the development and

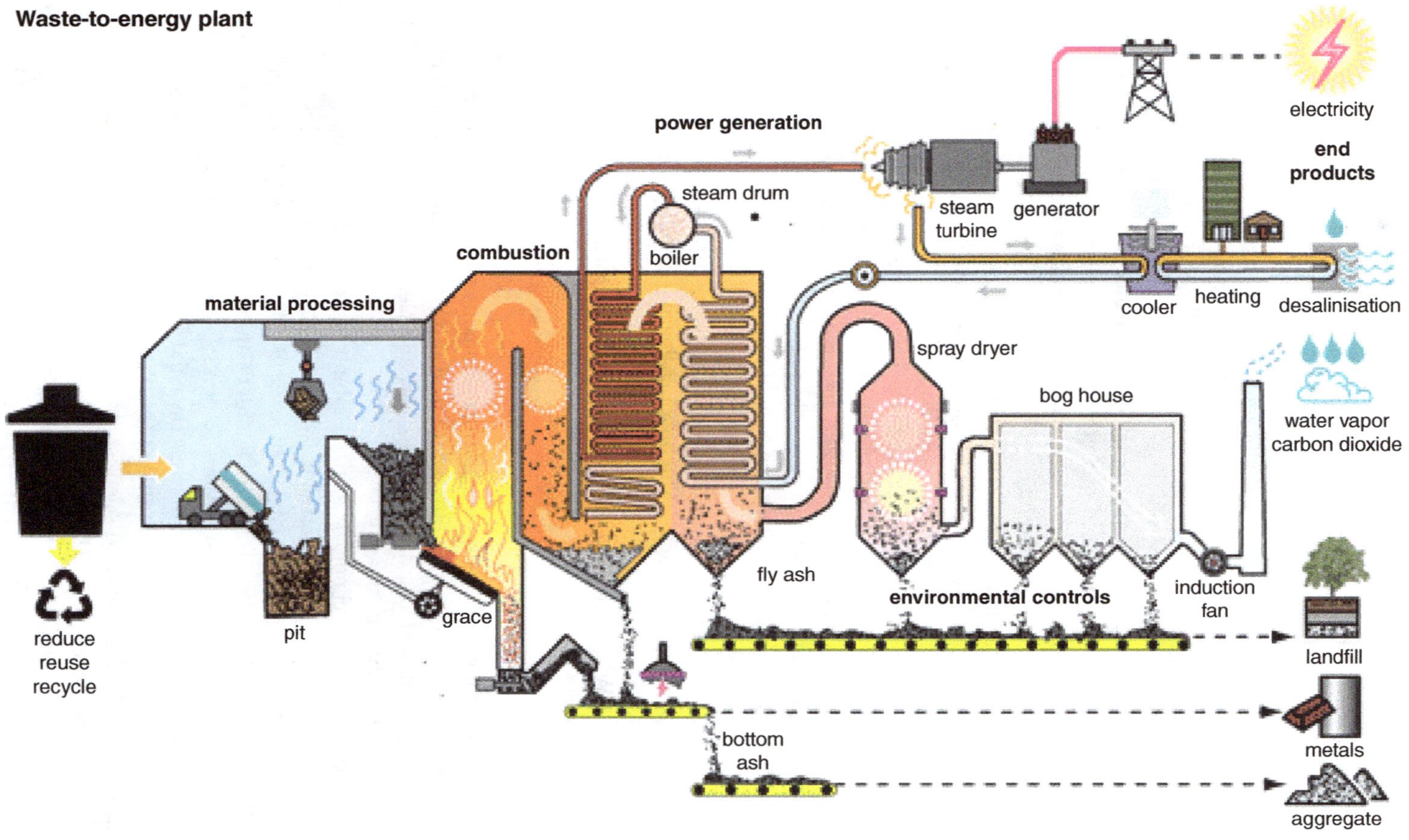

Fig. 10. Incineration process line to generate electricity.

commercialisation of innovative green solutions and working with foreign and local engineers to serve global needs for sustainable development.

The Research and Development (R&D) funding had been utilised to establish and grow public sector R&D centres in clean energy. The goal of its R&D efforts is to accelerate the cost reduction and efficiency improvements of solar panels, and improve the integration of solar technology with the grid. Solar Energy Research Institute of Singapore (SERIS) also runs the National Solar Repository, which tracks data from systems around the island in order to optimise solar systems in the tropics. CleanTech Park, located just outside of Nanyang Technological University in western Singapore, was designed and created by engineers in JTC to serve as a nexus for research, innovation, and commercialisation in clean technology that have embraced environmental sustainability.

Going beyond R&D, Singapore actively placed itself as a "living laboratory" for companies to testbed and commercialises innovative energy solutions, customised for the urbanised tropics. As a resource-constraint state, Singapore's main interest had been holistic energy management and efficiency, so as to improve our energy resilience, increase our energy options, and reduce our carbon footprint.

Given our land constraints, Singapore had been intensifying our competencies around rooftop-mounted solar systems that are optimised for the tropics. One such example is the Singapore American School in Woodlands that has installed a 1-megawatt-peak (MWp) solar panel system on its roof, one of the largest single-site installations here and the largest on a school campus in 2013 (Chua, 2013).

Another innovative approach is the ongoing floating photovoltaic pilot project on our reservoirs, as an alternative to rooftops. Solar cooling technology is another example where UWC, an educational campus, is testing the largest solar cooling system in the world today.

To cope with the growing clean energy sources as a limited land mass that is subjected to rapid fluctuating weather patterns, Singapore is also researching into highly novel control and optimisation platforms to address the challenges presented by intermittency of solar energy. Singapore is also piloting smart meters, building energy management systems, demand response, and electric vehicles as part of the overall national testbed in smart grids.

With the advance technology engineers built up over the decades, the once fuel-driven vehicles are slowly overtaken by electrical driven vehicles. The debut of electric cars is in the 2000s. Yet, engineers are still consistently working to improve the efficiency of these vehicles. Research and developments of hybrid cars have drew interest of many engineering institutions to develop their very own vehicles. Engineering undergraduate students from the National University of Singapore (NUS) built an electric car from scratch in 2013 (NUS, 2013). Bringing electrical cars to a commercial level, engineers and scientists from Nanyang Technological University (NTU) alongside with TUM CREATE — a research institution started by the Technische Universitat Munchen (or the Technical University of Munich) are developing taxis powered purely by electricity. The prototype, codenamed EVA was the brainchild of a team of engineers and scientists (Cheng, 2013).

Looking ahead, Asia will play an increasingly major role in the global sustainability landscape. Building on its strong competitive advantages, Singapore is already strategically positioned to do its part in driving green growth and helping companies implement such initiatives across Asia (Kiong, 2012).

References

Cheng, K. (21 November, 2013). *Today Online.* Retrieved 12 June, 2014, from http://www.todayonline.com/singapore/electric-taxi-prototype-launched-motor-show

Chua, G. (26 October, 2013). *Singapore American School goes big on solar power with 1MWp solar panel system* . Retrieved 13 June, 2014, from The Straits Times: http://www.straitstimes.com/breaking-news/singapore/story/singapore-american-school-goes-big-solar-power-1mwp-solar-panel-system

Chua, G. (23 Feb, 2013). *Wild Singapore.* Retrieved 30 June, 2014, from Singapore first coal-powered Plant here limits emission: http://wildsingaporenews.blogspot.sg/2013/02/singapore-first-coal-fired-power-plant.html#.U608CvmSySo

EMA. (2010). Retrieved 12 June, 2011, from Energy Market Authority: http://www.ema.gov.sg/page/32/id:65/

Energy 101. (13 May, 2013). Retrieved 29 June, 2014, from http://www.energyportal.sg/Energy101/types-of-energy.html

Energy. (22 August, 2013). Retrieved 11 June, 2014, from http://www.energyportal.sg/History/Timeline.html

Energy Market Authority. (2016). Singapore Energy Statistics 2016. 123.

Feng, Z., Chua, G., & Cheam, J. (8 Novemeber, 2011). *The Straits Time.* Retrieved 27 June, 2014, from http://erian.ntu.edu.sg/NewsandEvents/PublishingImages/MediaArticles/large/Hooking-Up-To-Region%27s-Powerhouses.jpg

Kiong, G. C. (Nov, 2012). *Future Ready Singapore.* Retrieved 13 June, 2014, from http://www.edb.gov.sg/content/edb/en/resources/downloads/articles/asias-clean-power-growth-frontier.html

NEA. (7 November, 2013). *National Environmental Agency.* Retrieved 17 June, 2014, from Solid Waste Management in Singapore: http://cgs.sg/wp-content/uploads/2013/11/Wkshop1d_NEA_Recycling.pdf

NUS. (17 December , 2013). Retrieved 12 June, 2014, from National Unversity of Singapore: http://www.eng.nus.edu.sg/ero/news/index.php?id=1620

Osmotic Power as an Alternative Energy Source. (n.d.). Retrieved 13 June, 2014, from http://www.pitt.edu/~eab84/EngineeringTrends.htm

Singapore Update. (29 Jun, 2004). Retrieved 6 Jun, 2014, from http://www.singaporeupdate.com/previous2004/previous300604_massiveblackouthitsmanypartsofsingapore.htm

SLNG. (2014). Retrieved 14 June, 2014, from Singapore LNG Cooperation: http://www.slng.com.sg/our-operations-terminal-facts-figures.html

SLNG. (2014). *Submission for IES prestigious engineering achievements award,* 3.

Taylor, A. (14 08, 2011). *The Atlantic.* Retrieved 10 06, 2014, from http://www.theatlantic.com/infocus/2011/08/world-war-ii-daring-raids-and-brutal-reprisals/100127/

Zero Waste SG. (1 April , 2013). Retrieved 17 June, 2014, from Singapore Waste Statistics 2012: http://www.zerowastesg.com/2013/04/01/singapore-waste-statistics-2012/

Chapter 4

Manufacturing

1. Introduction

Manufacturing is defined as the creation of large volume of products from factories with machineries operated by teams of people. Singapore had progressed over the past five decades from being a low-cost manufacturer of simple components by labour intensive operations to manufacturer of quality products by skill intensive workforce and then to a world leader in high-value products manufacturing with technology intensive equipment. With such progressive manufacturing record supported by a strong pool of engineers Singapore is currently moving from technology intensive to knowledge intensive industry of the future.

During the pre-war days of the 1920s, manufacturing activities in Singapore had been on small scale basis in individual businesses ran by artisans and skillful craftsman for things like spare-parts for ships engines, mining equipment, garments, processed foodstuff, etc for local consumption. There were also certain amount of manufacturing for the global market, examples like pineapple canning, latex and timber processing factories, etc, as South-East Asia where the world suppliers of such natural resources. One such sizable factory is the Ford Motor Works which was established in Singapore in 1926 at Anson road. This factory expanded and moved its car assembly operation to the Upper Bukit Timah factory in October 1941. In the 1950s, the years after World War II, factories were located at various locations all over the island, with majority of the sizable factories located along Upper Bukit Timah Road all the way towards the north zone in Woodlands. On top of the factories for pineapples, latex and timber, these newer factories were manufacturing products on large quantities, like motor car, connecting rods & car components, glass bottles, large sewer and water pipe, etc. During this period, there were only local craftsman and fitters working as technical workers in the factories with European engineers and technicians as their supervisors of which the majority being British and the Dutch.

After Singapore acquired its independence from the British after World War II, the new nation decided to embark on industrialisation as the route to economic growth and rolled out national plans in many fronts to achieve that goal. Our first industrialisation plan was coined by Dr Goh Keng Swee who become our first finance minister in 1959. He pointed out the we need to develop various industrial parks to welcome the global manufacturing giants to set up factories here so as to take up our abundance of manpower together with our well-positioned sea-port. The first plan was to build Jurong Industrial Estate which started construction in 1961 and went into operation in 1963 with 24 factories By 1968, there were more than 150 factories operating inside Jurong Industrial Estate which by then have a township of low cost homes and shopping units for the workers.

Thereafter, more industrialisation plans rolled on into the '70s with more industrial estates in Queenstown and Toa Payoh and more to come in the '80s and '90s in Ang Mo Kio, Yishun and other locations in Singapore. All these national development effort had attracted large scale world class manufacturing activities to our island and such national efforts will continue to keep our nation as an attractive location for advanced global manufacturing activities.

2. The Keys to Industrialisation

To create economic growth after World War II, the Singapore government needed to develop export-oriented industries by attracting global manufacturers to set-up factories in Singapore. To execute this the government set up the economic development board and various engineering schools to meet the need of manpower development.

2.1. *Economic Development Board (EDB)*

The Singapore Economic Development Board (EDB) was formed in 1961 to address the challenges of attracting foreign industrialists to Singapore. Besides preparing infrastructure like Jurong Industrial Estate and Jurong Port, EDB opened its first overseas centers in Hong Kong and New York to attract foreign investors. To advance industrial development, EDB promoted Singapore as a quick operations start-up location where factories were built in advance of demand with a highly skilled workforce readily available. With various types of products like scooters, garments, textiles, toys, wood products, hair wigs, etc manufactured in the early factories in Jurong Industrial Estates, EDB moved forward into 1967 and secured an S$6 million deal with Texas Instruments to manufacture semiconductors and integrated circuits in Kallang Basin. This major investment went into operation in July 1969 and marked the start of semiconductor industry in Singapore.

As the era of electronics product matured more EDB overseas centers opened in Zurich, Paris, Osaka and Houston between 1971 and 1976. With these additional offices, from the 1970s to 1980s, EDB pushed for more global electronic and semiconductor backend factories to set foot in Singapore and manufacturing became the largest sector in the economy, surpassing trade. Further to this, EDB planned to bring in the more related industries. As a result, Singapore's and South East Asia's first silicon wafer manufacture plant by STMicroelectronics opened in the early 1984.

Related to this, the manufacturing of personal computers, printed circuit boards, and disc drives were identified as important sunrise industries, and EDB worked to attract companies like apple, Motorola, Seagate, etc into Singapore from the 1980s to the 1990s.

As EDB attracted global factories into Singapore from the '60s through to the '90s, the supporting local engineering enterprises also became increasingly important to these MNCs. As such, EDB set up the Small Enterprise Bureau in 1986 and shaped a range of assistance schemes to help small local enterprises grow in tandem with the MNCs.

Moving towards the 2000s, EDB recognised the importance of innovative entrepreneurship and so launched the Start-up Enterprise Development Scheme (SEEDS) in 2001 (renamed SPRING SEEDS in 2007). In view of the wealth of investment opportunities to lower cost in China, India and the Asean region for lower cost, EDB set up numerous centres in these regions. It set up its Shanghai office in 2002, followed by the Beijing and Mumbai office in 2004 and office in Guangzhou in 2006. This resulted in Singapore's factories (both local enterprises and MNCs) having overseas infrastructure development projects such as the Batam and Bintan Industrial Estates in Indonesia, the Bangalore IT Park in India, the Vietnam-Singapore Industrial Park in Vietnam, the Wuxi-Singapore and Suzhou-Singapore Industrial Park in China.

For future industrial growth in high-value and high-technology, EDB further stimulates innovation, entrepreneurship and techno-preneurship. Thereby, attracted MNCs such as Hewlett-Packard (HP), Seagate, Johnson & Johnson, Procter & Gamble, Hyflux and other MNCs to set up innovation and leadership development centres in Singapore.

In 2011, EDB celebrated its 50th anniversary and one man who had been in much of all the above was Mr Philip Yeo who was the Executive Chairman, Economic Development Board (EDB) from January 1986 to January 2001 and Executive Co-Chairman, EDB from February 2001 to January 2006. Mr Yeo redirected EDB's focus from the established fields to new areas of business. These included: internationally exportable services; developing high-tech industries like electronics cum semiconductors, information technology, aerospace, biomedical sciences, chemicals, etc. He also encouraged Singaporean companies to make direct investments abroad and take up pioneering role in entrepreneurship and techno-preneurship for future industrial economic growth.

2.2. *Pioneer status*

In the 1970s–1980s, to lure foreign investors' participation into Singapore Manufacturing Industry, Pioneer Status was awarded to companies which manufacture goods that had never been manufactured in Singapore. Besides being allocated sites for setting-up factories, one of the benefits of receiving such a status includes tax exemption over five years of operation in Singapore (Championing Manufacturing,

2012). Now, to encourage the growth of high-tech/high value-added manufacturing and services industries, Pioneer Incentive includes full corporate tax exemption on qualifying profits for up to 15 years (EnterpriseOne, 2010).

2.3. *Singapore Manufacturing Federation (SMF)*

The Singapore Manufacturers' Association (SMA) was established in 1932 with 17 local manufactures as founding members with the aim of representing the interests of local manufacturers during the British rule. During that period, there was an urgent need to diversify the economic base of Singapore since the main export commodities namely rubber and tin were badly affected following the Great Depression of 1929.

In the period following independence, the SMA contributed to the economic growth of the country's manufacturing sector by spearheading "Made-in-Singapore" brand of products overseas through trade missions as well as at international fairs and exhibitions. The association's first annual large-scale trade exhibition held in 1970 involved over 100 manufacturers, and sought to demonstrate the progress of industrialisation in Singapore through the range of goods on show. The promotion of "Made in Singapore" brand was well received by the audiences and trade missions to promote Singapore-made products overseas became part of SMA's regular activities.

In the 1970s, SMA members categorised various industry groups to better cater to their different needs. The groups included: Food and Beverage, Textile, Furniture, Chemical, Building, Metal, Rubber and Plastic, Electronics and Electrical, Companies in Jurong. The SMA, renamed in 2003 as the Singapore Manufacturing Federation (SMF), has grown from 17 founding members to over 3,000 members.

Manufacturing industries took a change in 1978, turning from a labour intensive production to a high-value added industry such as electronics, semiconductor and petrol chemical. Over the years, the association has emphasised higher productivity and standards in manufacturing through quality control, skills training and innovation of which a full spectrum of engineering manpower in the industries were leading and participating in full force. This leads to improvement of the quality of products and export competitiveness of Singapore goods.

As Singapore industries prospered in the 1980s, Singapore became an expensive place to do business. Hence, government pushed local enterprises overseas to establish their footprint in order to maintain their competitiveness. In alignment with the Government's advice, SMA sought to assist local businesses to establish their regional presence and overseas operations with lower costs, thus helping to grow the external wing of the Singapore economy.

From 1997 to 2012, manufacturing industries was redefined to include all services and industries that contribute to the manufacturing value chain. During this period from 1997 to 2012, the Singapore manufacturing industries have grown a comprehensive spectrum of high-tech, high-values manufacturers. These are both local enterprices as well as MNCs involved in designing an manufacturing electronics, computers, diskdrives, tooling, machineries, automated systems, etc with the local pool of manufacturing engineering manpower.

The National Productivity Board (NPB) was established as a statutory board under the Ministry of Labour on 12 May 1972. The board was tasked to promote productivity; assist companies in raising productivity through training of management and supervisors.

The most visible outcomes of the NPB's work were the introduction of Japanese labour-management practices and productivity concepts. In September 1981, the board launch the National Productivity Council and the Productivity Movement. The implementation of the Total Productivity Approach, which is a conceptual framework that considers productivity from the pillars of manpower, capital (hardware) and environment helps the factories to review themselves for improvement in productivity.

Some of the better known programmes initiated by the Productivity Movement included the formation of Work Improvement Teams and Quality Control Circles where engineers and management take the lead to work with teams of the technical workforce to solve problems in the production floor leading to improved quality and productivity. Teamy the productivity bee mascot (Fig. 1) and the introduction of the National Productivity and Singapore Quality awards all helped to push the nation's industries to achieve national productivity goals with an international recognition of high quality in Singapore manufactured products.

While WIT and QCC were running across the industries, concurrently, the Singapore Institute of Standards and Industrial Research (SISIR) was set up in 1973 to help our manufacturers test and validate the quality level of our locally manufactured product against international quality standards. In April 1996, the NPB merged with the Singapore Institute of Standards and Industrial Research (SISIR) to form the Singapore Productivity and Standards Board (PSB).

Fig. 1. Teamy the Productivity Bee icon by National Productivity Board (1982 to 1999).

2.4. Training engineering manpower for industry

2.4.1. Skillful technicians

During the pre-wars days, the local industrial manpower were artisans and skillful craftsman trained by their masters through apprenticeship arrangements through family or clans associations. The engineers in the factories then were mainly found in British and a few foreign corporations and were mainly from Britain or other European countries. The British government had trade schools established in 1929 at Scott Road area offering primary school leavers 2-year courses in mechanical and electrical fitting, plumbing and construction skills. These moved to Balestier area in the 1940, but due to World War II, proper classes really began in 1948. This trade school is reorganised into a vocational institute in 1963 as the Singapore Vocational Institute (SVI). After the war and with independence from the British in the 1960s and 1970s, vocational training was managed by two government bodies, namely the Adult Education Board (AEB) and the Industrial Training Board (ITB). As the country move on with industrial growth, these 2 boards merged in 1967 to become the Vocational and Industrial Training Board (VITB) as the sole national authority on technical training to produce suitable technical manpower for the factories.

As our industrial growth towards the 1980s brought in foreign MNCs in the high-tech areas, the requirement of technical manpower moves a grade higher, requiring the entrees into the VITB to be at secondary school level instead of primary level. This shaped the 1992 effort for a new branding of VITB into ITE the Institute of Technical Education which holistically handle the education of technical manpower for the industry till this day.

2.4.2. Specialised technologists

In the late 1960s, during the start-up phase of Singapore industrialisation plans, the government noted early that there is a need to bring in expertise to train our labour for the specific skills that the foreign investors' factories needed. Therefore, in 1979 EDB set-up 3 special skills training schools with Foreign MNCs namely Tata, Rollei and Philips government training centres. These centres run apprenticeship programmes that trained our people specifically in tool and die-making, precision metal machining and precision mechanics for 2 years in the centre followed with another 2-years attachment to their respective employers' factories either overseas or locally.

In the mid-1970s, the rapid development of Singapore's economy from labour-intensive industries to technology and knowledge-intensive industries resulted in an extremely tight labour market for technologists to fill in jobs in the MNCs. Therefore Singapore requested for foreign governments to assist technically and financially with the establishment of *institutes of technology*, taking in trainees with *A* level qualifications and maintaining close ties with technology leaders from the industry — serving as a conduit for technology and know-how transfer. In 1978, a proposal to establish the German-Singapore Institute (GSI) was put forward by EDB and this institute commenced operations in February 1982, complementing the other training for vocational and technical manpower. This was followed by the French-Singapore Institute (FSI) that commences in 1984 and specialises in the training of technologists in the electro-technical engineering fields, modelled after the then École Supérieure d'Ingénieurs en Électronique et Électrotechnique (ESIEE) of Paris. Together with the Japanese-Singapore Institute of Software Technology (JSIST) that started in Singapore polytechnic in 1980, these establishments attempted to fill an essential gap in vocational and technical training, providing critical manpower to new and merging industries from the '80s to the '90s.

In 1993, government decided to put all these institutes into mainstream education system with GSI and FSI combined into the Precision Engineering Institute at Nanyang Polytechnic and with JSIST evolving into School of Media and Info-Communications Technology (SMIT) currently at Singapore Polytechnic.

2.4.3. Engineers and scientists

Singapore's Ministry of Education (MOE) was established in 1955 after the war. At that time, the policies in education are focused to be aligned with Malayan nationalism concerns and there was no need to train engineers. As the local government was concerned that the education in Singapore is too academic to support the coming industrialisation plans, MOE in 1956, established two technical secondary schools and a secondary commercial school as well as a Joint Advisory Council for Apprenticeship Training. By 1959, MOE was able to offer 1,820 places in vocational and technical schools.

Besides vocational and technical training, the Singapore government in 1956 also started professional engineering education in the then University of Malaya (UM) at Bukit Timah campus. By 1958, the department of engineering was elevated to full faculty status and move back to Kuala Lumpur, Malaya. In the same year, MOE opened the Singapore Polytechnic as a constituent college of UM offering 4-year professional diploma course. In 1962, the University of Malaya reorganised into 2 campus, with the engineering faculty of the University of Singapore set-up at the Singapore Polytechnic at Prince Edward road. In 1968, the first batch of Singapore Polytechnic graduates received their Bachelor of Engineering degree from the University of Singapore. By 1969, the Faculty of Engineering was officially constituted under the University of Singapore with three departments, namely Civil, Electrical and Mechanical. To further meet the needs for manpower to run the growing number of factories, the department of Industrial and Systems Engineering was established in the Faculty of Engineering in 1972. With the growth of student population, the Faculty of Engineering moved to Kent Ridge campus in 1977 and eventually renamed as National University of Singapore in 1980, serving the nation's engineering manpower needs as the nation industrialisation effort moves along with global trends towards high-tech era.

In the mid 1980s, our manufacturing industries moved from labour intensive towards high-tech era with advanced semiconductor, telecommunication and disk-drive technologies booming all the way into the '90s. In order to fulfill the need for graduates and postgraduates manpower for these high-tech industries, the National University of Singapore set up the Postgraduate School of Engineering offering Masters and PhD courses in 1990s. This was further followed in 1995 with the introduction of B-Tech part-time degree courses so as to groom more engineers from the diploma pool of employed manpower.

3. Industry Progression

Various types of manufacturing industries entered Singapore at different time over the last 50 years of nation-building. This chapter traces Singapore engineering community effort to match with the timeline of increasing manufacturing activities in order to advance the manufacturing competitiveness of Singapore's economy. Singapore manufacturing industries went through a few transformations as we kept pace with global manufacturing advancements. In the 1950s–70s, manufacturing activities went through from labour-intensive to skills-intensive phase. Further growth in the manufacturing industries during the 1980s to 2000s, was contributed by moving from skills intensive into technology intensive activities involving design and development of products and automation. Since the 2000s onwards, the manufacturing industries had moved onto adopt IOT and with widespread proliferation of advanced automation. In parallel, there is increased in research and development activities in industry leading towards the future of knowledge based industry.

3.1. *Early industries*

Jurong Industrial Estate was developed in the late '50s and by 1962, the first factory established in Jurong Industrial Estate was National Iron and Steel Mills, now known as NatSteel. At the start in 1963, Jurong Industrial Estate housed 24 factories and with continuous development by JTC, there were over 150 factories established there by 1968.

National Iron and Steel Mills had been a pioneer and key partner in Singapore's nation building efforts. The Singapore plant is the only local steel mill with an integrated upstream and downstream operation, where steel is manufactured through recycling scrap. Products from NatSteel, like re-bars are widely used in most construction projects islandwide, from the iconic Changi International Airport to public housing that is the trademark of Singapore's landscape.

Other types of factories in the '60s were making scooters, electrical console boxes, biscuits, chocolates, metal cans, etc, These industries required large number of low cost unskilled labour as operators or assembly man as they are just making traditional products with relatively few machinery managed by the foreign investors'

Fig. 2. Hume industries (now Hong Leong Building Materials Pte Ltd) Pipe Manufacturing factory in Singapore, 1965.

engineers and there were no need for local engineers besides some skillful craftsman for machinery maintenance (Fig. 2).

By 1972, Matsushita Refrigeration Industries (S) Private Limited set foot in Singapore. This is the first Panasonic factory located here to manufactures refrigeration compressors, electrical components and cast iron parts. Subsequently known as Panasonic Refrigeration Devices Singapore Pte Ltd. (PRDS), it was Singapore's first Panasonic under Panasonic Corporation. Today it operates as a subsidiary of Panasonic Asia Pacific Pte Ltd and it houses the division's international HQ outside of Japan and manufactures its full product range, contributing 10% of all household refrigeration compressors to global markets.

3.2. Electronics industry

Singapore's electronics manufacturing started in the late 1960s to the early '70s following the establishment of Jurong Industrial Estate. Back then, labour-intensive industries were encouraged to invest in Singapore to take up the delicate wiring and soldering jobs required in electronic products as our labour force consisted on many women besides man.

In 1965, Setron Limited produced the first locally assembled television sets, allowing Singapore to be recognised as the only TV assembly plant in Southeast Asia. By the 1970s, Setron Limited had become a household brand in Singapore, known for the durability and reliability of its black & white television sets. From producing just 400 TV sets a month in 1966, the company ramped up production to 2,500 sets a month in 1973. In 1986, the company became a subsidiary of Sony Corporation Japan and its name was changed to Sony Singapore.

Philips started its factory operation in the '70s in Singapore. This company started in 1951 in Singapore as a trading firm with a four staff that imported and sold lighting products, radios and gramophones. With this humble beginning, Philips introduced Singapore to its world of manufacturing lighting and consumer electronic products. Today, Singapore is the regional headquarters and competence centres for Philips' three sectors — Lighting, Healthcare and Consumer Lifestyle. The company also undertakes various global product development, training and production activities in Singapore.

HP Singapore was opened in April of 1970. The operation was established to manufacture core memories for HP211X computers. HP went to great pains to emphasise that Singapore was chosen not because of low cost labour but because the women there were more suited than those elsewhere to do the intricate stringing of the tiny doughnuts onto the fine copper wires in core memory production. HP was able to reduce the cost of its core memories by about thirty percent then. By manufacturing in-house, HP also reduced its reliance on outside suppliers. Within months of opening, HP Singapore also began manufacturing low cost diodes for the HP Associates Division.

By September of 1973, HP had 1,800 employees in Singapore, still operating out of the top two floors (50,000 square feet) leased from the Singaporean government. HP Singapore was the fastest growing HP entity ever. Employment at HP Singapore did not reach 2,000 until early 1980.

As the electronics industry continue to grow through to the 1980s and early 1990s, low-skill labour-intensive industries were phased out and being replaced by a mature electronics industry with both MNCs and local home-grown companies. These now form the core of the high technology industries requiring skill and talent where manufacturing engineers of various disciplines were important manpower running the show.

Various brands namely Philips, Hewlett Packard, Panasonic, Apple, Motorola, Singatronic, Compact, etc together with Singaporean Innovative company Creative allowed Singapore's electronic industry to grow and become a vital node in the global electronics market. Making random picks from any electronics gadget today garners a good chance that parts from the gadget was designed or made in Singapore.

A globally known local company Creative Technology Limited was established on 1 July 1981 in Singapore by Sim Wong Hoo. This 35-year-old company, brought to the world the sound blaster electronic card that brought sound to the computer world. This company has about 3,000 employees continuously working on its range of personal digital entertainment products such as portable and wireless speakers and audio MP3/MP4 players. It continues to develop the Sound Blaster processors, which were used in computers and game consoles such as the PS3. In 2012, Creative Technology announced the production of Creative HanZpad platform. Differing from others in the market, the Creative HanZpad platform enhanced with features for Chinese characters, targets China's vast tablet computer markets, especially the education market at an opportune time when its government aims to transform the conventional education system to one that embraces the latest in digital technologies.

With electronic companies like Setron, Philips, HP, etc operating in Singapore, they bring with them the technology to mould plastic parts and factory management techniques from Japanese, European and US to be imparted to our local engineers. In the early years, the local manpower are only operating the machines in the factories but by the '70s, these MNCs needed the local supporting industries to make spare-parts for their machine tooling. From this started the precision engineering industry in Singapore where high-end precision tooling like plastic mould and stamping die for various industries began to be design and manufactured by the local workforce comprising of engineering manpower trained in local polytechnics, technical institutions and later universities.

3.3. *Precision engineering industry*

Singapore's precision engineering activities began in the 1970s to support the MNCs. This industry originally consisted of local component fabricators in the '70s and now consisted of modules manufacturers and equipment design cum manufacturing companies. Today, there are some 2,700 companies, ranging from small and medium enterprises (SMEs) to large multinational corporations (MNCs), in the precision engineering sector. Singapore also plays host to the headquarters and R&D functions of many of these companies.

As the backbone of manufacturing, the precision engineering sector is a key enabler for Singapore's high-value manufacturing. This industry contributes to 10% of Singapore's manufacturing output and employs close to a quarter of the total manufacturing workforce. With its highly specialised skill sets, precision engineering is an integral part of the manufacturing chain for all kinds of sophisticated products, ranging from the smallest semiconductor wafer to cutting-edge medical devices, advanced aerospace components and the large drill bits used in oil exploration.

In this industry, the components manufacturers are mainly local small and medium size companies who are making components for global brands in various industry sectors like machine-tools, semiconductor equipment, electronic equipment, hard-disk-drive, consumer electronics, industrial machinery and equipment, etc. Each of these companies have their own process skill-sets that is required for the various types of components that needed to be manufactured and together they form a local eco-system of capabilities that is valuable to the continuous growth of various sectors of our local manufacturing industry (Fig. 3). Some of the early companies in the '70s and '80s were CKE Manufacturing, Sunny Metal & Engineering, Amtek, Seksun, Fong Lee Metal, Metal Component Engineering, Sanwa Plastic Industry, Sunningdale Tech, Yeakin Plastic Industry, OMNI Mould, Fuyu Mould, MC Packaging, Stamping Industries, etc.

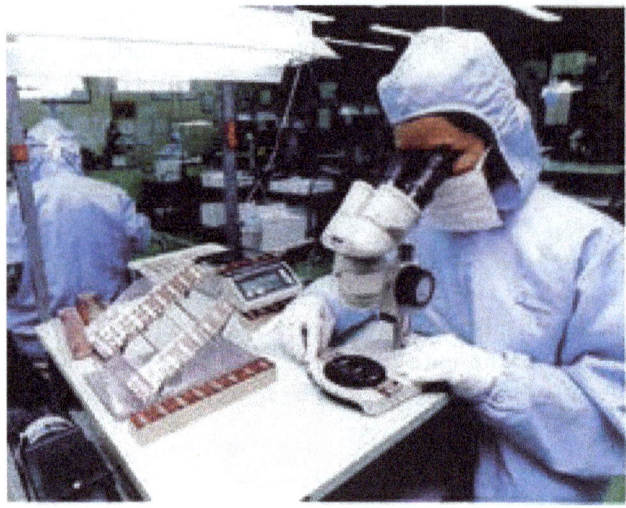

Fig. 3. Mitsui High-tec Singapore making precision stamped leadframes for the semiconductor industry in the 1990s.

As for the group of modules and complete system manufacturers, these are larger multi-national companies both foreign and local owned who take on contract manufacturing jobs at higher assembly levels from the global brands and they are normally named as contract manufacturers. These companies normally do not manufacture the components themselves but out-source these to their partner suppliers and they concentrate to hone their skill-sets in assembly and test of customers' product. Starting with local contract manufactures who had grown into the region, like Fuyu, Venture, etc. Singapore also attracted global contract manufacturers like Flextronics, Jabil, etc.

Besides the contract manufactures operating here, there are also the global brands of precision engineering companies who are manufacturing their own product series here in Singapore. LeBlond Makino and Okamoto are precision engineering companies which began operations in the 1970s. For Makino Singapore, precision craftsman and engineers were involved in the manufacturing of vertical milling machines, wire and RAM EDMs machine tool, while Okamoto craftsman and engineers manufactures grinding machines for the world.

Started in the '70s were various local precision engineering companies in injection moulding and metal stamping and many had grown to take up wider role in the contract manufacturing of precision modules and sub-systems for their customer. One of them is Fuyu Mould which was occupied with the fabrications of injection moulds and the manufacturing of plastic injection parts. Into the '90s, Fuyu Corporation grew and extended their operations to include various sub-assembly processes for moulds, precision modules and products to become one of the largest one-stop suppliers in precision mould fabrication, injection moulding and assembly in the region. Many other precision engineering companies which started in the '70s and '80s follow similar growth paths to cover a wider scope in precision engineering manufacturing activities for the global market in semiconductor, medical component, optics, aerospace industries, etc.

As for the many MNCs in precision engineering, the local engineering manpower had been able to keep up with the new skills needed in their continuous growth in Singapore. One good example is a division company of Sony Electronics, Sony Precision Engineering Center (Singapore) was established on 3 February 1987. It manufactures precision components: optical pick-ups, magnetic heads (Hi-8, DVC), digital 8, drum (digital data storage and spindle motors), SMT, power supply, solution business. In year 2005, the company had a staff strength of 350 people. Since its establishment in 1987, SPEC has grown in importance to become Sony's center for production technology in Asia.

3.4. Semiconductor industry

Singapore's semiconductor industry began quite humbly in the 1960s, with early Assembly & Test (A&T) activities. A&T activities are generally regarded as the back-end stages of semiconductor production, in contrast to front-end activities such as wafer fabrication and integrated circuit (IC) design.

With Texas Instruments setting up transistor assembly factory in Kallang in 1969, other semiconductor backend players such as Fairchild Semiconductor, SGS (now known as STMicroelectronics), National Semiconductor, Panasonic, etc followed earnestly. When these companies came into Singapore in the late '60s, these early days were still the age of labour intensive manufacturing activities like IC assembly and testing. It was not until the late '70s that skill intensive activities were brought in and the industry needed engineering trained graduates from the polytechnics and university to take up roles in this semiconductor backend manufacturing industry as process engineers, equipment engineers, failure analysis engineers, operation engineers, maintenance, facilities, industrial engineers, etc.

Engineers from the mid-80s onwards focused their effort to bring up the productivity of these factories through the latest automation technology together with manufacturing engineering methods. With the help of the National Productivity Board, engineers led total quality management through WIT (Work Improvement Teams) and QCC (Quality Control Circles). Engineers also adopted global techniques and concepts like cost of ownership management through OEE measurements and preventive maintenance management, etc. Till today, semiconductor backend factories like Infineon, UTAC, Statschippac, Qualcom, Skyworks, Lumileds, etc have their engineers focusing in productivity improvement programmes with enhanced tool kits and latest technology from the global automation and IT industry.

By the early '80s, Singapore changed its emphasis in foreign investment, moving away from the skill intensive back-end semiconductor industry towards asset-rich and design-based wafer fabrication front-end of the semiconductor sector.

This transforms the sector into one that manufactures higher value-added products. By 1981, SGS started to set up its first wafer fabrication plant and IC design centre in Singapore. The Singapore engineers and technical specialists were sent for training in Italy and the fab in Ang Mo Kio started to produce its first wafers in 1984. From the late '80s onwards, many other wafer fabrication plants began to start-up in other parts of Singapore. In 1987, Singapore Technologies Engineering started the Chartered Semiconductor wafer fab in Woodlands (from 2010 sold to Global Foundries). In 1997, NXP joined with TSMC to create SSMC fab which started operation in year 2000 in Pasir Ris wafer fab industry park. In the same park, UMC from Taiwan started its fab in 2004. In the late 2000s, a collaboration between Intel and Micron in NAND Flash technology created Singapore's first 300mm NAND Flash wafer fabrication plant in Feb 2007 under IM Flash Singapore (IMFS).

In this front-end semicon industry, majority of engineers are in the process plants of the wafer fabs improving productivity of the fab operations through the latest innovative precision equipment technology. In most of the wafer fabs, wafers are transported automatically in a Front Opening Universal Pod (FOUP) that has a capacity of 25 wafers. These wafer FOUP movements are executed using Automated Material Handling System (AMHS) in 300 mm fab or by Automatic Guided Vehicle (AGV) system in 200 mm fabs. In all these automation systems, equipment-to-equipment communication technology are being employed with real-time optimisation in scheduling and tracking to further increase operating efficiency. There are also IC design engineers working in the wafer fab affiliated fabless IC design houses producing innovative IC designs to meet the ever-advancing needs of new type of electronic gadgets.

Today, Singapore alone is home to 14 front-end FABs, 20 packaging and test operations, and more than 40 IC design companies, including nine of the top ten fabless chipmakers. In addition, Micron Semiconductor Asia Pte Ltd worked with A*Star Data Storage Institute to develop a next generation non-volatile memory storage.

3.5. Hard disk drive industry

With a strong electronic and semiconductor industry made up of MNCs and a pool of supporting local precision engineering and automated equipment companies, the hard disk drive industry moved into Singapore in the early '80s. This comes at a time when Singapore industry starts its transition from low skill to skill intensive industry. At that time Singapore had thousands of precision engineering manpower trained in the mid-70s from the Rollei training centre looking for jobs when Rollei left in early '80s. This pull factor attracted many hard disk drive makers like Seagate, Maxtor, Miniscribe, Conner, Western Digital, IBM, etc. From 1986 to 1996 during the high time in this industry, Singapore accounted for 50% of hard disk drives shipped around the world and employed around 80,000 people.

Seagate being the first hard disk manufacturer to setup factory in Singapore, had employed local management and engineers. During the era from 1986 to 1996, the whole hard disk industry in Singapore saw the local engineers and their management implemented various manufacturing system engineering and management concepts like lean manufacturing, six sigma, lean design, etc in order to keep manufacturing cost competitive for their hard disk drives for many years.

As hard disk drive is a device in any computer, and when the computer industry drifted to notebooks, tablets and handphones, competition drove the hard disk drive companies to keep changing their product models and designs. This ever shortening of the hard disk product life cycles posted to the automated production lines in the factories a huge challenge for fast turn-around time to meet production schedules. While majority of Seagate competitors were dropping out of competition, the engineers in local equipment companies took up the challenge to design robots into production lines so as to enable short conversion time for production lines to produce different models (Fig 4). This changed the entire concept of equipment design for automated production lines for hard disk drive and enabled fast change in the models of hard disk drive product for the global computer industry. From the late 1990s to early 2000, due to this and other factors like rising labour and land cost, majority of the hard disk manufacturers had dropped out of the competition, leaving Seagate, Western Digital, Maxtor and Hitachi Global Storage Technology in Singapore by 2005. By the end of 2009, Seagate closed its Ang Mo Kio plant and with it, all the hard disk manufacturers are out of Singapore.

Fig. 4.   Matsushita disk-drive manufacturing with Pana Robot in the 1990s.

While hard disk drive manufacturing is moving elsewhere, companies in data storage devices like Seagate moves into new data storage applications areas including mobile PCs, local backup storage devices, digital video recorders and new opportunities such as network attached storage solutions. One such company, Showa Denko HD Singapore was incorporated in 2002 in Singapore to manufacture HD media. Their plant at Pioneer Crescent started operation in Jan 2003 and subsequently had two plant expansion exercises, in 2006 and 2010 to cater to growing demands for HD media. Similarly, in 2004, Seagate opened its reading media operation in Woodlands in Singapore. In 2015, Seagate incorporated their new technology research and development centre in our One-North research hub along Ayer Rajah Crescent. Named *The Shugart*, after company founder Alan Shugart, the R&D centre will focus on the development of 2.5-inch small form-factor hard drives, hybrid drives, firmware, software and technologies. The future of data storage devices and applications will be exciting as the personal mobility gadgets are coming out in many creative ways and cloud storage is still a much needed function.

3.6. *High-tech equipment industry*

In the last 40 years since gaining independence, per capita income in Singapore had increased an average of 6.4% per year from 1965 to 2000. Yet that growth has been driven largely by manufacturing rather than knowledge-based industries. Singapore realised that if we stay this way, we cannot keep up competitiveness as our regional countries are catching up fast in skill intensive manufacturing at very much lower cost. Therefore Singapore have to make a change to support high-value knowledge-based industries. Since the '70s, Singapore had our local industrial manpower trained in the electronics, semiconductor and precision component manufacturing industries, making the local manufacturing sector a good platform to take on high-tech equipment manufacturing, design and research and development activities.

At the end of 1990s, global brands in electronics and semicon high-tech equipment products starts to set foot in Singapore for their manufacturing activities, utilising the relatively low cost well-trained engineering manpower that is needed to perform the hi-tech assembly and testing tasks. One of the early companies, KLA-Tencor, started operations in Serangoon North industrial estate in 1997, first sourcing for precision components from our local precision fabrication shops and follow with assembling optical and other modules for their high end wafer inspection equipment. In 1998, Applied Materials a global leader in nano-manufacturing technology for semicon and solar wafer equipment, setup a training centre in Singapore. This is followed by a new facility in Changi North industrial park in 2009 which serve as operation center for Applied Material's global purchasing, sales, manufacturing, engineering and financial groups to support the regional Asian semicon chip industry and the rapidly growing solar markets in India and China. This new operation will help Applied Material to achieve reductions in material cost and cycle time while expanding their global supply chain capabilities and will also continue to support Singapore's strong semicon chip manufacturing industry. In 2000, K&S set up assembly operations in the same industrial estate to manufacture their wire bonders for semicon industry.

With the top-notch global semiconductor equipment players moving here for equipment manufacturing operations in the '90s, our EDB also invited the MNCs in scientific and medical equipment products to join in this high-tech equipment sector here. In 2007, AB SCIEX, an international leader in life science analytical technologies, established Singapore as its sole global manufacturing site for all instruments. In June 2012, Life Technologies, the second largest player in the life science tools industry, announced that it had established a Global Instrument Centre of Excellence in Singapore to address strong customer demand around the world.

In 2005, Edwards Lifesciences, the world's largest tissue heart valve company, has its only Asian heart valve manufacturing facility located in Singapore. In March 2011, Medtronic the world medical technology leader opened its first Pacemaker and Leads manufacturing facility in Singapore. Medtronic continues to build on this presence, and in August 2013, Medtronic officially opened its global Centre of Excellence (CoE) for Business Model Innovation in Singapore. This new facility is responsible for designing, testing and scaling new business models for the rapidly growing emerging markets across Asia. Besides MNCs, in 1989, we also have a locally started company manufacturing critical care products for the biomedical industry. Biosensors went on to be a Singapore-listed company and established a manufacturing facility here in 2014 for its flagship drug eluting stent, known as the BioMatrix.

Besides the high-tech equipment companies from electronics, semicon, scientific and biomedical industries, Singapore have a local company Hyflux incorporated in year 2000 as a high-tech water desalination equipment company disrupting the water treatment industry. Hyflux manufactures high-tech membranes that are compacted into filtration modules that are the crucial process modules for water desalination plants built worldwide.

With the new entry and growth of these high-tech equipment industry players together with the growth of existing CNC machine companies like Makino, Mazak and Okamoto; our local manufacturing engineering manpower and precision fabricators supply chain will have to move towards knowledge-based industry with jobs involving design, development and research. Many of these high-tech equipment companies had set up R&D section in their local factories here or with the local research institutes and universities. As this high-tech equipment companies setup more R&D activities, the current group of engineering manpower together with the future young engineers from the tertiary institutions will be given all the opportunities to develop and grow further so as to have a seat in the future high-tech world market.

3.7. Process industry

The process industries are those industries where the primary manufacturing processes are either continuous, or occur on a batch of materials that is indistinguishable as opposed to discrete and countable units. For example, a food processing company making sauce may make the sauce in a continuous, uninterrupted flow from receipt of ingredients through packaging. In Singapore, the process industry covers food, petroleum, petrochemical, specialty chemicals and pharmaceutical products.

Besides the process and equipment maintenance engineers required to operate these processing plants, a large portion of our local engineering manpower for these process industries are in the Engineering Service Providers (ESPs) that support in the areas of plant construction and plant maintenance, involving conduit piping, pressure vessels, boilers, etc. The association of process industry (ASPRI) was established in 1997 to help the process industry in infrastructure, internationalisation, management practices, mechanisation, workers' capabilities, performance metrics and turnaround scheduling, etc. It is estimated that there is about 500 ESP with different specific engineering skill areas serving the local process industries.

3.7.1. Food industry

In the 1950s and earlier, Singapore food manufacturers were predominantly domestic-oriented and comprised a majority of family-run small and medium enterprises (SMEs) and a number of global names that started in the early 1900s. After Singapore independence in the '60s, the food & beverage manufacturing industries that flourished then were companies like Fraser & Neave in 1972 with automated processing and bottling of aerated drinks in River Valley area and the installation of beer and stout processing plants by Malayan Brewery (late known as Asia Pacific Brewery) in Jurong Industrial Estate. Such were examples at the beginning of automated processing in Singapore's food and beverage manufacturing industry and it laid the fertile ground for the development of process industry manpower.

By the 1980s, some local companies such as the Super Group (Thong Siek Food Industry) had already ventured into the global market with more product varieties utilizing automated processing plants in Tuas Industrial estate that enabled them to roll out their products at faster rates that reduces cost. Global names like Nestle who had been in Singapore since 1912 also moved towards automated process plant since the '80s. Currently, Nestle operates one major factory, in Jurong, which is the company's largest malt extract manufacturing plant in the world and it also has a world-class infant formula manufacturing plant in Tuas. In the 1997, Barry Callebaut also started automated chocolate product processing lines in Senoko industrial area after acquiring Van Houten chocolate factory built in the '60s in the Queenstown industrial area. Similarly, local Amoy Canning which first set up a factory in Singapore in 1951 had moved to a new factory in 1994. The new factory in Jurong industrial estate was streamlined for high capacity production and equipped with modern food processing and packing machinery.

To support the food & beverage industry to compete in international market, The Food Innovation and Resource Centre (FIRC) was setup in 2007 to provide food enterprises with technical expertise in new product and process development including packaging, shelf life evaluation and market testing and to provide training to the industrial manpower for this industry. Besides their fundamental engineering knowledge, the process and maintenance engineers in these process plants have to know about Food Science and Technology, have to implement Food Hygiene & Food Safety Management System and to implement international practices like GMP and HACCP for Food Manufacturing.

To further move this industry, the Singapore government set up the Clinical Nutrition Research Centre (CNRC) in 2013, which is a joint initiative between the Singapore Institute for Clinical Sciences of A*STAR and the National University Health System. This center specialises in basic and translational human nutrition research involving studies across the life cycle. These include investigation of the impact of micro-and macro-nutrient intake on human physiology and understanding the role of food structure on human nutrition.

3.7.2. *Pharmaceutical and biotechnology industry*

Besides the food & beverage process industry, Singapore brought in and grow the high value-added pharmaceutical sector of process industry. In 1972, GlaxoSmithKline (GSK) started a single amoxicillin plant in Singapore and has expanded since 1982 into a network of three process plants manufacturing API (Active Pharmaceutical Ingredients).

In 2000, the Singapore Government announced a big push to develop the biomedical sector, setting up the Tuas Biomedical Park (TBP) which is a 360-hectare stretch of ready-prepared and specifically-zoned land for pharmaceutical and biologics manufacturing. The TBP comes with all essential infrastructure, such as roads, power lines, telecommunication lines, sewer pipes and water and gas supplies done by the ESPs. Also on standby are third parties providing utilities such as steam, natural gas, chilled water and waste treatment services. With the estate's *plug-and-play* design, pharmaceutical, biologics, medical device and other biomedical companies can set up manufacturing operations with minimal lead time. At the same time the government also setup research and development support infrastructure for this industry at Biopolis (at one-north), which has a state-of-the-art infrastructure for the life sciences. Biopolis co-locates public sector research institutes with corporate labs and is designed to foster a collaborative culture among the research institutions and industry organisations.

With good infrastructure and capable manpower, a number of leading biopharmaceutical companies like Abbott, GlaxoSmithKline, Lonza, MSD, Novartis, Pfizer and Sanofi-Aventis have chosen to make Singapore their global manufacturing location to manufacture a wide range of active pharmaceutical ingredients (APIs), biologics and nutritionals. To date, all pharmaceutical manufacturing facilities that had started operations had received validation from international regulators such as the US Food and Drug Administration (FDA) and the European Medicines Agency (EMEA).

The pharmaceutical and biotechnology manufacturing sector in Singapore is currently supported by a growing base of more than 4,800 skilled engineers and technicians. Similar to the food & beverage industry engineers and technicians, the manpower in this sector also have to gain additional knowledge in pharmaceutical and biotechnology and to implement international practices like GMP and HACCP for their production plants. A good case of food as well as nutritional product is milk powder manufactured by Abbott's S$450 million nutritional powder manufacturing plant in Singapore. This process plant is Abbot's first major capital investment in Asia and its largest nutritional investment to date. This S$450 million state-of-the-art facility was completed in April 2008 to meet the increased demand for Abbott Nutritional Products in the region. The Singapore facility manufactures Abbott brands like Similac Advance® infant formula, Gain® growing-up milk for older babies and toddlers, Pediasure® and Grow®.

According to Data Monitor, Singapore was the third fastest growing nation globally in the export of pharmaceutical products from 2000 to 2010. The bio-pharmaceutical industry expanded by more than 30% in 2011 and contributed about S$22.8 billion in output and over 6,000 jobs. The type of engineers includes Automation Engineer, Facilities/Utilities Engineer, Maintenance Engineer and scientific personnel like Analytical Chemist, Biotechnologist, Scientist, Molecular and Cellular Biology Scientist, Analytical Biochemist, etc. As such, Singapore remains committed to developing a manpower base that is ready for biotechnology and pharmaceutical manufacturing requirements of the future.

3.7.3. Oil & gas and chemical industry

Singapore is the region's premier hub for oil & gas, a sector that contributed almost 5% to Singapore's gross domestic product in 2007. Singapore is the world's top three export refining centres and the oil industry accounts for 5 per cent of Singapore's gross domestic product. The oil industry is not a standalone industry. Refining has been the catalyst for the chemical industry, providing advantaged feedstock as well as other spin offs including oil & gas equipment and oil rig manufacturing sectors.

In the 1890s, Singapore oil & gas industry started with trading of mainly kerosene and lubricants under the Mobil Oil brand name. After World War II, by 1963, Mobil went into refining activities with Singapore's first refinery, reaching 18,000 barrel-per-day (bpd) capacity in 1966 at Pioneer Road in Jurong industrial estate. By the late sixties, Esso also entered the refining business, resulting in a 90,000 bpd refinery built on Pulau Ayer Chawan (now called Jurong Island), which in 1970 produced 296,000 bpd.

In 1973, Singapore Petroleum & Chemical Company (Private) Limited, started a refinery on Pulau Merlimau with refining capacity of 70,000 bpd. This is followed by Singapore and South-east Asia first joint processing refinery which was incorporated on 11th January 1979 under Singapore Refining Company Pte. Ltd. (SRC). This is a joint venture by Singapore Petroleum Company Pte. Ltd, British Petroleum Company Limited and Caltex Petroleum Corporation. Further to this, SRC commissioned in 1983 the SRC Visbreaker Complex and Catalytic Reformer Complex and in 1986 the Hydrocracker Complex. SRC continued its expansion with the commissioning of a Residue Catalytic Cracker in 1995 (Fig 5).

In the 1980s and 1990s, Mobil and Esso continued to upgrade their refining capacities as well as add more downstream petrochemicals and lubricant facilities. In 1999, the merger of Exxon Corporation and Mobil Corporation named ExxonMobil started to build a single new state-of-the-art US$2 billion petrochemical plant on Jurong Island which is fully integrated with existing refining and chemical operations. As the industry grows, ExxonMobil's Singapore Chemical Plant Expansion opened in 2014 is the single largest manufacturing investment in Singapore's history. This petrochemical complex is ExxonMobil Corporation's largest integrated chemical and refining site.

For the process industry, in the few decades from the 1960s to the 2010s, all these refinery, chemical and pharmaceutical facilities investments were made possible by engineering manpower from the infrastructure and construction sectors of industry. Jurong Island, is an integrated energy and chemical hub, to facilitate trade and manufacturing activities. The construction of Jurong Rock Cavern, a massive underground facility, give Singapore a million cubic meters of space for the storage of crude oil, condensates and naphtha. Other infrastructure facilities built like the Biotech & Pharma Park in Tuas industry estate adds to the competitiveness for growth of the process industry. Together with the large number of engineering service providers (ESP), the operations of these processing plants are manned and maintained by a large pool of process, instrumentation, chemical engineers, biomedical scientists and technologists working round the clock on Jurong Island and the nearby Jurong industrial estate. As the industry constantly upgrades capabilities to operate state-of-the-art technologies, companies can tap on a highly-skilled workforce capable of managing high-end complex manufacturing and research projects.

4. Engineering Contribution and Innovation

Generally, three industry clusters in the electronic sector namely semiconductor (both front-end wafer fabrication and backend chip packaging), hard-disk drives, printed circuit board assembly stand out among the electronics manufacturers in Singapore. In fact, electronics is the bedrock of the Singapore manufacturing sector till today, contributing 5.2% to the country's Gross Domestic Product (GDP) in 2012.

In the '80s and '90s, locally we do not have much engineering R&D platforms except small corners of our universities lab. However, in line with the progress of science and technology, our local engineers do learn and progress along through

Fig. 5. The second naphtha cracker facility being developed on Jurong Island in the mid-1990s.

their workspace, as we were well-prepared by our polytechnics, and universities and innovation in technology goes on in parallel with what is happening in the USA, Europe and Japan. While some of the MNCs do have R&D activities, the bulk of innovative work back then happens mainly in small private start-up companies by whose founder engineers had experience and exposure in MNCs.

In the mid-80s, when the engineers in the semiconductor backend factories saw thousands of operators working hard to improve the yield of their output for good productivity numbers, they realised that without automated equipment to handle the minute wires in the integrated microchips there is no improvement possible. Coupled with the advancement of electronics and computers technology, these local mechanical and electronics engineers started to design and build automated equipment with precision for high-speed and high-volume production using less operators.

Concurrently, engineers with entrepreneur mindset moved out from the MNCs to set up support industries like machine-shops, equipment design and manufacturing companies, etc. These players formed the basis for the precision engineering industry and a whole new equipment manufacturing industry started.

In the late '70s there were various SME machine-shops making spares parts and tooling assemblies for the MNCs in the various industry sectors. In the early '80s, one of these companies, Microfits, began to design precision moulds for the final packaging of integrated circuits (or IC chips) in the semicon backend industry. Their innovation in gang-pot mould design was awarded patent rights and improved the moulding operation productivity of the semicon backend operations thereby opening a whole sector of precision mould design and manufacturing industry service provider for the semicon industry in the '80s. From this, the company Advanced System Automation was incorporated to create automation for moulding machines as the productivity of the Microfits mould is too fast for the operators to cope.

In the semicon backend sector, local engineers are the ones who designed, developed and manufactured higher speed higher volume production lines as compared to those in the United States and Europe where there was no high-volume production. One such early company is I.C.Equipment who designed and developed bulk tube handlers and automated cap and base loaders, etc that were first in the world. These automated equipment lifted the productivity of the semiconductor backend factories in multiple folds (bulk tube handlers changed the operator to machine ratio from 1 to 1 to 1 to 4 and the cap and based loaders change the throughput of production from 2,000 units per hour to 10,000 units per hour or reduce operators from 4pax to 1pax). These inventions not only helped the Singapore factories but also many semiconductor backend factories all over Asia.

In the component manufacturing sector we had local SME machine-shop like Polymicro whose CNC engineers started to fabricate precision components for the hard-disk drive factories in the late '80s and thereby opening a chapter in contract manufacturing activities for components used in high volume consumer products built locally by the large MNCs. With this metal component contract manufacturing trend, the local engineering entrepreneurs went on to take on the non-metal components contract manufacturing activities like building PCBs and PCBAs for the MNCs like Apple, Motorola, etc. This activities started with PCI and follow with the notable Venture Corp and others.

In the '80s, within the PCB and electronic industry grown a group of engineers who started to design and assembled their own personal computers. One prominent case is the local Cubic 99 computer that was launched in 1984. From this emerged our inventor Mr Sim Hong Hoo who in 1988 invented the sound-card for personal computer. Before Mr Sim's invention appeared, computers were just dumb machines used in offices and corporations. Mr Sim's idea of having sound in a computer changed the way the world accepted computers. With sound from computers, we opened the doors of computers to incorporate multi-media content for many purposes, thereby drawing in the mass population to use computers. With the advancement of computer industry from desktop computer to personal computer to mobile PC to mobile tablet and handphones of today, Mr Sim's innovation of a digital sound card for computers had made a very big part in the global history of computer technology.

From the '90s onwards, the booming hard-disk drive industry co-existed together with the semicon microchips industry within a small 20 km zone in Singapore. Through these activities, in the late '90s, evolved another local engineer Mr Henn Tan who invented the USB thumb-drive storage (Fig. 6) that was launched in year 2000. This disruptive product totally removed the floppy disk from the computer industry globally. Till this date, the thumb-drive product is still a much loved personal mobile storage device that is used widely around the world. The company TREK2000 who had launched the thumb-drive continued to push out innovations like the Flucard and I-Ball till date.

Fig. 6. A USB thumb-drive from Trek 2000.

In the past 50 years, local engineers and their team had worked diligently in their respective engineering roles to bring about good quality products manufactured in Singapore through the various companies MNCs and SMEs. On top of that, the eco-system in the local industry had brought out numerous innovations that were disruptive global products from our engineering people. We cannot possibly list all in this chapter but just to record some notable cases here. As the global manufacturing scene is getting much more challenging and competitive in the future, we hope that the spirit to be innovative in the face of adversity will be strengthened further by our future generations of engineers.

5. Future Venture in Manufacturing Activities

Engineering being the fundamentally important element in modern society will continue to become increasingly indispensable into the 21st century, especially for manufacturing where it is more difficult to find and more costly for human labour. Building upon the strong cohort of engineers and engineering eco-system that was already built up locally over the last 50 years, our future engineers will have a good ramp to stand on while facing the various new challenges and opportunities in manufacturing.

In manufacturing, the next few decades will see high level of implementation of mechanisation integrated together with the latest digital tools and internet connectivity technology (Fig. 7). In the '90s when computer integrated manufacturing was known, the level of computer and internet connectivity technology was still not agile enough for the majority of manufacturing problems. Today, after more than 30 years of advancement, the digital world has very high speed computers,

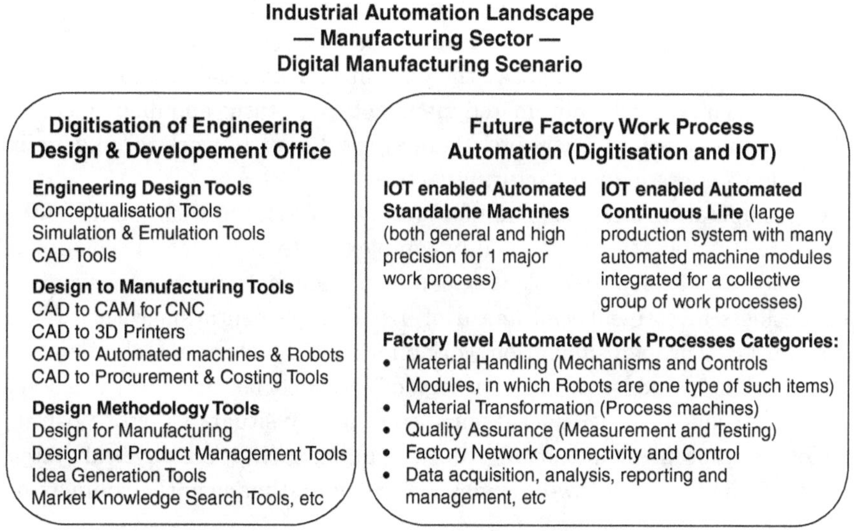

Industrial Automation Landscape
— Manufacturing Sector —
Digital Manufacturing Scenario

Digitisation of Engineering Design & Developement Office

Engineering Design Tools
Conceptualisation Tools
Simulation & Emulation Tools
CAD Tools

Design to Manufacturing Tools
CAD to CAM for CNC
CAD to 3D Printers
CAD to Automated machines & Robots
CAD to Procurement & Costing Tools

Design Methodology Tools
Design for Manufacturing
Design and Product Management Tools
Idea Generation Tools
Market Knowledge Search Tools, etc

Future Factory Work Process Automation (Digitisation and IOT)

IOT enabled Automated Standalone Machines (both general and high precision for 1 major work process)

IOT enabled Automated Continuous Line (large production system with many automated machine modules integrated for a collective group of work processes)

Factory level Automated Work Processes Categories:
- Material Handling (Mechanisms and Controls Modules, in which Robots are one type of such items)
- Material Transformation (Process machines)
- Quality Assurance (Measurement and Testing)
- Factory Network Connectivity and Control
- Data acquisition, analysis, reporting and management, etc

Fig. 7. Manufacturing sector diagram.

very minute sensors, highly mobile devices and reliable high speed wireless network connectivity to enable the design and development of highly automated equipment and systems for majority of manufacturers.

A large portion of Singapore manufacturing scene is mainly in light-industry, with operations churning our high mix high volume type to high mix low volume type of products. Some of these operations will benefit from highly automated lights-off operation model and some will require highly agile operation model or a Singaporean model to survive the global competition. To this end, our local engineers will have to be innovative to integrate all the available advance technological tools in the world into equipment and system designs that our manufacturing businesses can adopt.

5.1. *Digital tools and IOT technology*

The abundance of digital tools like computers, mobile tablets, hand-held scanners, cameras, RFID tags, etc can help manufacturers to plan schedules, run operations, track work-in-progress, track machine status, etc. Having real-time high-level visibility of their manufacturing operations will improve the time to market respond of the manufacturers to the fast changing global demand for manufactured goods. To achieve good implementation and maintenance of such fully automated manufacturing systems, we need our engineers to understand manufacturing operations and design the optimised digital systems that suit the business of each factory. Besides the factory operational systems, there are machineries in factories that have to be highly integrated with all the advanced digital devices and IOT technology to make themselves into fully automated equipment and production lines. In particular, wireless technology together with advance sensors technology had enabled the autonomous vehicles or robots to be roaming in factory floors to carry out material handling activities. But such implementation still need engineering and R&D work to make this into mass adoption. At the current moment, only some parts of some factories had achieved fully automated manufacturing status and many factories are still in manual operation. Therefore, there are still a lot of work to be done in the next decade to upgrade manufacturing factories.

Besides operating factories, manufacturers are also faced with the tasks of planning the factory operations with highly complex customer demand scenarios. Now there are advanced software tools that can help manufacturers plan their manufacturing business investment well ahead of a real implementation with a relatively low cost outlay. Through simulation work on a virtual factory model, manufacturer can test out the response of their factory or intended factory to future scenarios like resource changes, order fluctuations, contingency situations, etc. With digital technology advancement in software tools, building such cyber-physical systems of factories will be the most low-cost means to plan and manage the complex manufacturing operations in the near future.

Since the '80s, we saw the advancement in computer-aided design tools from 2D to 3D tools and now into advanced engineering simulation tools. These technologies in design tools had enabled the design process from the sketch of ideas all the way to simulations of physical process outcome of the design into a series of virtual tasks that can be done by designers. With this change in design process, design engineers can build multiple proto-types in cyber-space, thereby saving time and cost in the building of multiple physical proto-types of their innovations. Our future engineers have to learn and adopt the design and simulation tools to enable themselves to be agile and competitive in the new manufacturing world where better and faster design customisation of customer product needs is a global trend to be sought after.

Further to this, with the advancement in 3D printers, many components can be 3D printed and this make it a straight through process from design to printing the components. While the simulation tools are powerful tools for an improved product and equipment development process in terms of saving prototype cost and time to manufacture the final product, the adoption of 3D printing technology for widespread use in manufacturing is still too early to tell.

5.2. *Big data and artificial intelligence*

In the years to come, with the advancement in technology in big data analysis, local engineers will be expected to utilise big data analysis technology to derive useful insights and critical information for the competitive edge of the manufacturing plant they are serving. One area of application is in prognosis where big data analysis can predict process and or machine degradation ahead of time, thereby enabling remedial planning and actions to be initiated before a costly breakdown can happen.

The second area of application is for the equipment manufacturers and factory users of equipment to have in-depth diagnosis of yield problems in their processes so that the root cause of quality issues can be identified accurately thereby cutting short the route to correct solution for quality products. Such big data analysis technology for manufacturing had currently started its in-route to our local companies and many engineers are actively trying to learn and implement it for some good cause in their organisations.

With the implementation of big data analysis technology in equipment and systems, we will expect to have intelligent equipment and systems populating our factories and our life in the next decades. By then further work on artificial intelligence will probably lead us to deploy an intuitive human-like robot as a co-worker in the factory.

To achieve future success, engineers of all disciplines must have an open mind to think and integrate their knowledge in multi-disciplinary teams to solve our society's problems as a united group without boundaries.

References

Championing Manufacturing. (2012). In *Partner in Nation Building*.
EnterpriseOne. (2010). Retrieved 2014, from Pioneer Incentive: http://www.enterpriseone.gov.sg/Government%20Assistance/Tax%20Incentives/Product%20Development%20and%20Innovation/gp_edb_PC-M_PC-S.aspx
Jurong GRC. (n.d.). Retrieved 11 July, 2014, from Tribute to Dr Goh Keng Swee, the "Father of Jurong": http://www.juronggrc.sg/goh_Keng_swee

Chapter 5

Buildings & Infrastructure

1. Introduction — From a Low-Rise to High-Rise Nation

From slums to high-rise buildings, Singapore physical landscape has changed tremendously over the past decades. Along with the building progress is the advancement of technology. Sophisticated technologies have allowed the duration of construction to cut down by twofolds and improved the construction quality despite the rising height. This has allowed a space-restricted island to meet growing demands at an efficient pace.

Yet, the challenges of constructing a building are more than the tip of the iceberg. Some of the biggest civil engineering challenges facing Singapore include:

- space constraints, thus leading to potential underground expansion development;
- variation in soft soil conditions that pose challenges to foundation construction and
- design of sustainable green buildings by architects that poses structurally challenging buildings for construction.

Ironically, these challenges allow civil engineers to showcase their creativity and resolve the issues on hand. Hence, structurally and aesthetically, Singapore has managed to clinch a number of international awards in these areas.

Therefore, this chapter is to commemorate the building and construction heroes behind Singapore's nation-building.

1.1. *Singapore building history*

1956

Kampong Lorong Buangkok

It is one of the surviving Kampong in Singapore till date. Built in 1956, using materials such as zinc for roof and walls (Fig. 1), Kampong Lorong Buangkok housed 28 families, 18 Chinese and 10 Malays. Electricity and water supply were accessible to the families in 1963, 2 years before independence.

Fig. 1.   Houses in Kampong Lorong Buangkok.

1965

Approximately 70% of Singapore's household lived in badly overcrowded conditions, and a third of the remaining population squatted on the city fringe.

1.2. *The start of a housing revolution*

When Singapore was still under the British colonial government, public housings for the masses were planned and built by the Singapore Improvement Trust (SIT) (Fig. 2). However, SIT was slow in churning out new houses for the growing population, and majority of the population still lived in unfavorable living conditions.

Shortly after gaining self-independence, the newly elected government, the People's Action Party, passed the Housing and Development Act of 1960, and the Housing & Development Board (HDB) replaced SIT.

HDB was then led by the late Lim Kim San (1916–2006) who was their chairman and also Singapore's Minister for National Development.

Recognising the need to quickly provide the masses with quality housing, the HDB began constructing massive high-rise flats that can hold many units. Places such as Queenstown and Tanglin Halt were among the first few estates that HDB started developing.

Fig. 2. Dakota Crescent: Flats developed by SIT in the 1950s. Some of them lasted until the late 2010s.

Between late 1960s and early 1970s, Taman Jurong was earmarked for residential housing development in order to provide convenience for industrial workers that work within the Jurong vicinity. As such, the Jurong Town Corporation (JTC) was formed to oversee the construction of the Taman Jurong estate.

Moving toward the end of 1970s, more than 50% of Singapore's population were housed in HDB flat, a remarkable improvement from the 8.8% that lived in flats built by SIT (Chew, 2009). Eventually, HDB extended their development plans to areas further from city center such as Lim Chu Kang, Sembawang, and Clementi.

After many years of experience in building, HDB started introducing new block designs and better flats (Fig. 3) to attract the increasingly wealthy middle-income group.

Today, 82% of the population lives in HDB flats. To cater for the ever-increasing population, more flats will have to be built. Staying true to its mission of providing affordable housing and vibrant living spaces, HDB has continued to innovate and improve flat designs over the years, breaking new grounds with projects such as the award-winning Pinnacle @ Duxton (Fig. 4).

2. Construction Technique and Features in the Past and Present

2.1. *Improving the productivity of our construction industry*

With increasing construction demand and limited resources in Singapore, there will be an urgent need for the construction industry to raise the productivity and hence efficiency of the industry.

Fig. 3. HDB has introduced a variety of flat designs over the years, as can be seen from this photo taken in Bishan.

Then-Senior Minister of State for National Development and Trade and Industry Lee Yi Shyan said in 2013 that Singapore's built environment sector's productivity had to achieve at least 30% improvement by 2020 to closely match those in the advanced economies.

Mr Lee added that companies are using up to two times the manpower to complete the same project than the best international benchmarks. To improve productivity, he added that the sector can learn from the infocomms and manufacturing industries in terms of seamless industry integration and up-front planning. The engineers have therefore come up with several innovative and creative solutions to constantly improve the productivity of our construction industry.

2.1.1. *Use of prefabrication techniques*

One interesting solution employed by our building engineers in improving construction productivity is through the use of prefabrication technique. Prefabrication is a two-stage construction method in which prefabricated structure components are produced in a factory or workshop in the first stage and then assembled into buildings at the site where they are to be erected.

Fig. 4. Pinnacle @ Duxton, the tallest public housing building in Singapore.

Prefabrication is able to increase the productivity as the construction time on the site is greatly shortened since the structural work on site is reduced to only the construction of foundations and erecting the prefabricated components. Second, low-skilled and lesser number of manpower will be needed since the precast units are manufactured in the factories under factory conditions that make use of lesser manpower as compared to traditional on-site construction.

The statutory board responsible for public housing in Singapore, HDB, has succeeded in developing and refining its own concoction of semi-precast system that is able to achieve high quality and construction productivity in the development of these high-rise buildings. Being the pioneer and leader in prefabrication technology in Singapore, HDB has since maintained the precast implementation level at about 70% for each project. Besides improving construction productivity, there are also other benefits that can be reaped through prefabrication:

- Quantities of materials required are reduced since precast units are usually assembled using cranes or other handling and lifting machines; scaffolding and formwork are largely eliminated. This also provides for a safer working environment.
- Production of a large series of standard precast units allows the use of machines and automation to reduce manual labour.
- Better quality of the products is obtained as precast units manufactured under factory conditions are constantly under quality control and better working environment.
- Construction can proceed regardless of the weather conditions on site.

2.1.2. Increased use of technology

Another area that our engineers have helped to increase the productivity is the greater use of technology to promote efficiency and effectiveness. One such example is the use of Building Information Modeling (BIM) on our construction projects (Fig. 5). BIM is a digital representation of physical and functional characteristics of a facility. It is also a shared knowledge resource for information about a facility

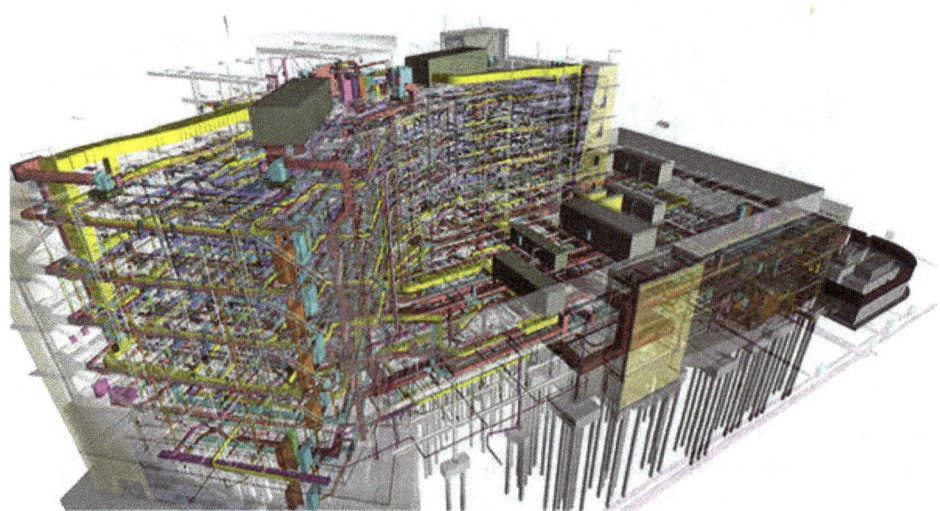

Fig. 5. BIM can be used to model both the internal and external features of a facility being constructed.

forming a reliable basis for decisions during its life cycle, defined as existing from earliest conception to demolition.

The key of BIM to improving construction productivity lies in its ability to allow project partners of different discipline to develop solutions together, better manage the project risk, and enhance decision-making. This will help to reduce the administrative time of passing the project through the different partners and also reduce the errors due to miscommunications between these partners. For instance, general contractors and prefabricators, such as concrete precasters, can utilise BIM to share critical information during the construction process and agree on solutions early during the planning stage which will in terms translate into seamless and more productive off-site manufacturing and on-site assembly workflow.

With BIM, a three-dimensional model of a project and drawings can be shared among the professionals, allowing them to analyse and resolve potential design clashes before construction begins. BIM can facilitate better teamwork among professionals, helping to reduce unnecessary reworks when the project is being constructed.

2.1.3. Alternative materials for construction

Another step that engineers have taken to improve the productivity of the construction sector is the switch from traditional construction materials of sand and concrete to steel construction. Many advantages can be reaped from the abovementioned switch. Besides helping to reduce the depletion of natural resources and to enhance the resilience of our construction industry, improving building sustainability, steel construction also improves productivity.

This is because it reduces the time spent for formwork, reinforcement work, concrete strengthening, and removal of formwork, speeding up construction as compared to that of traditional concrete-based construction method. In addition, steel components are generally slimmer and can support a longer span. A significant saving in foundation is possible as steel has a higher strength-to-weight ratio as compared to bulky concrete structures.

Although this switch is more costly, it, however, brings about many other benefits. Challenging structures can be constructed using steel. The present Supreme Court is a good illustration on how steel helps in enhancing engineering creativity (Fig. 6).

Another alternative material being employed is the high impact resistance dry wall. It has gained popularity among the residential and commercial developments. Dry wall has many advantages over the conventional wet construction method (i.e. using brick wall or concrete wall). Dry wall construction is cleaner, lighter, faster, and can be easily handled by fewer workers.

Projects such as St Regis Residences, The Sail @ Marina Bay, Residences @ Evelyn, The Pier @ Robertson are just some examples of those that have adopted dry walls to replace conventional brick walls. More projects are joining this trend as the industry players gradually recognise the benefits in using dry wall system.

Fig. 6. The Supreme Court building constructed in 2002 used steel construction with composite steel decks for its superstructure.

2.1.4. *Automation in construction*

Looking ahead, Singapore is also planning to employ a higher level of automation in construction such as the use of robots to further improve the productivity of our construction sector. The Building and Construction Authority (BCA) has looked into technological projects such as mini-piling robot, self-climbing painting robot, and concrete finishing robot.

2.2. *Making our buildings barrier-free*

In the early days of public housing when HDB had to quickly resolve a housing shortage crisis, the special needs of the elderly and disabled had been admittedly subsumed together with the needs of the other segments of the population. But as early as mid-70s, HDB initiated a study on barrier-free design features to the building and its immediate surroundings with the first pilot project incorporated in Ang Mo Kio Town Centre.

Creating a barrier-free environment in public housing, homes to 86% of the population have been a concern of HDB. This takes on added importance in the face of a rapidly aging population. It is anticipated that in the 2030s, almost one-fifth of the population will be aged 65 and above and most of these elderly will be living in public housing apartments.

A study on the barrier-free features to the building and its immediate surroundings was initiated in the early 1970s with the first pilot project incorporated in the Ang Mo Kio town center. By 1985, HDB had introduced barrier-free design

Fig. 7. Ramps that facilitate mobility up elevated ground (left) and Braille plate implemented on the buttons of HDB lifts to aid the visually-impaired (right).

at all levels of the neighborhood, precinct and apartment block in its public housing towns.

At the neighbourhood level, the focus is on building a network of barrier-free and vehicular-free walkways to connect each precedent to the amenities within the neighborhood. This greatly enables the elderly as well as the disabled to have safe and easy access to the commercial and recreational facilities within the neighborhood.

At the block level, a major improvement is the lift accessibility. In the earlier days, block designs had allowed the lifts to stop only at intermediate levels. Since then, lifts are designed to stop at every floor of new public housing blocks, and this enhances the accessibility of the elderly and the disabled. Furthermore, lift cars are provided with features such as handrails, lower lift call buttons, and doors wide enough for wheelchairs. In addition to these, voice synthesiser and Braille plates for the visually disabled have been incorporated in all lifts (Fig. 7).

In 1990s, HDB recognised the need to upgrade those older public housing estates that had been built shortly after Singapore's independence. The lift upgrading programme was one of the projects done by HDB to improve the flat's accessibility and personal convenience. Apart from that, several other projects are carried out to make the internal of the flats, the common areas, and the external built environment more elderly-friendly and accessible.

3. Greening our City

Singapore is a densely built-up urban environment with limited natural resources and land space. As such, green buildings are vital to the nation's sustainability. Come 2030, Singapore will be much greener than today when as much as 80% of the buildings is certified green. With green buildings, we can now achieve efficiency in energy and water. Integrated with more green spaces and the use of ecofriendly materials for construction, Singapore can move toward sustainable development.

In 2005, BCA kickstarted its drive in greening Singapore's physical landscape by launching the BCA Green Mark. BCA Green Mark is a rating system established to evaluate the environmental impact of a building and recognise its sustainability performance. The benefits of Green Mark buildings include cost-saving resulting from efficient use of key resources such as energy and water, leading to lower operational and maintenance costs. Other less tangible benefits include enhanced occupant productivity and health due to good indoor environmental quality.

The introduction of the certification program was a bold initiative to move Singapore's building and construction industry toward environment-friendly buildings by providing a standard benchmark and guideline for the industry to follow.

Buildings are assessed under the following criteria:

- Energy efficiency
- Water efficiency
- Environmental protection
- Indoor environmental protection
- Other green features

3.1. Building strategies

As a small island city-state with resource constraints, we have to use our land, energy, water, and all available resources prudently to achieve a sustainable development, meeting the demands of people now and in the future.

To green our city, building owners adopted technologies and approaches such as integrated and passive design, energy and water efficient strategies, and building greenery. In this aspect, our architects, consultants, and engineers have been contributing by studying and implementing technologies suitable for our buildings so as to minimise harm to our environment.

3.1.1. Building siting, massing, and orientation

The initial site planning of a building is important to achieve a green and high-performing building. Building siting, massing, and orientation of buildings have an effect on resource consumption of the building. In terms of passive design, site planning is the first step in minimising the building energy demand, providing daylight, natural ventilation, shading, and thermal comfort. It is a complex challenge to put all the features and design together, and it requires skillful judgment by engineers and architects.

Solar heat gains from direct solar radiation lead to increased cooling load and hence energy consumption. In spaces with natural ventilation, solar heat gains heat up spaces such that they typically become thermally uncomfortable to occupants. To minimise solar heat gains, it is critical to optimise the orientation and massing of a building specific to its location. Certain orientations (East and West, for example) are more exposed to the sun and therefore have a greater solar heat gain. On the other hand, the massing of a building could provide shade for itself and other blocks to mitigate solar heat gains.

Natural Ventilation

One strategy toward reducing the energy demand is to maximise the amount of naturally ventilated space because natural ventilation requires less energy than air-conditioned areas. In Singapore, the wind usually blows from the north to northeast or south to southeast depending on the monsoon season. Although Singapore generally has low wind speeds, the velocities achieved are enough to provide comfort to spaces with the help of optimised design to allow natural ventilation.

Daylighting

While minimising solar heat gain is important, it is also important to take advantage of and harness the natural daylight for spaces. This strategy helps to reduce the need for artificial lighting which requires a significant energy demand. Bringing in daylight via window openings at appropriate heights, skylights, and/or atrium spaces is an effective strategy that will affect massing and orientation decisions.

Optimised Orientation

In Singapore, the sun is almost directly overhead throughout the year. East and West orientations receive the most solar exposure, and hence most solar heat gains. According to the Sun Path, both North and South orientations also receive high solar exposure for a portion of the year.

To overcome the problem of high solar radiation, the massing and orientation of the building must be well planned to minimise East and West facing facades (Fig. 8).

Materials

Heat island effects are experienced at higher local temperatures due to the surrounding environment. To mitigate heat island effect which is experienced in Singapore, architects and engineers have developed numerous strategies, ranging from roofing and hardscape materials to amount of green space and shade provided.

Pervious materials for surfaces are used to allow infiltration during rain. Some pervious concretes are concrete, asphalts, pavers, and open-grid paving materials. Also, it is encouraged to use light-coloured or reflective materials without creating unwanted glare to neighborhood buildings. Materials must have high solar reflectance values in order to absorb significantly less solar radiation than dark-coloured materials.

Greenery

Maximising the amount of greenery on site is one strategy to provide shade and reflect solar infrared radiation and keep building areas cool, improving the thermal

OPTIMISE SOLAR ORIENTATION:
NO OPENINGS ON THE EAST OR WEST FAÇADE

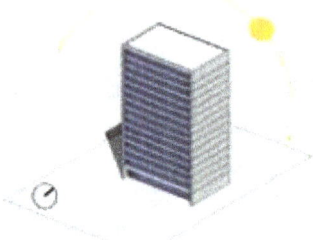

On East and West facing façades, the number and size of openings should be minimised. It is also better to have more opaque wall area to mitigate solar heat gain. A good strategy is to plan stairs, elevators, bathrooms, or other "noncore" spaces toward this orientation.

SELF-SHADING: BLOCK A SHADES
BLOCK B EAST FAÇADE

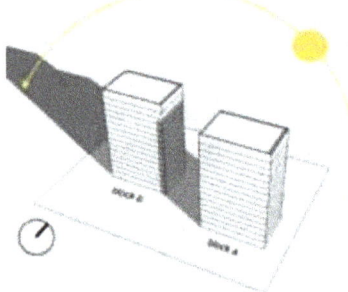

For projects with multiple buildings, massing to shade each other can be used, particularly for East and West facing façades.

Fig. 8. Minimise solar heat gain by reducing openings on the East or West façade (top) and using massing to shade adjacent buildings (above).

comfort of the area. Wherever possible, greenery should be provided at ground level, in planters, indoors, and on roof areas (green roof) inclusive of vertical greening system.

Rainwater Harvesting

Rainwater is harvested at the rooftop of each residential block to help lower the usage of potable water at the common area.

3.1.2. *Building envelop*

Building envelop separates the interior and exterior environments of a building. It serves as an outer shell to protect the indoor environment. Building envelop design is a specialised area of architectural and engineering practices that draws from all areas of building science and indoor climate control.

Glass Properties

Properties of glass must be selected based on its performance properties for both thermal and visual. Glass performance is differentiated based on a number of properties such as visible light transmittance and shading coefficient.

Glass properties have significant impact toward reducing the cooling load. Any number of measures can be undertaken to improve glazing performance, each with their own advantages and disadvantages.

Shading Devices

Another strategy to reduce heat gain is to install sunshade. An external sunshade can often be used as a design feature, with its primary function to reduce solar heat gain. Its secondary functions would be to reduce solar glare, provide rain protection for opening windows, and to serve as part of a maintenance strategy. All in all, sunshade design must be informed by solar geometry and Sun Path studies to ensure its effectiveness.

Greenery

Vegetation reduces solar heat gain and improves microclimate within any interior or exterior space. The incorporation of plants into buildings is a popular feature in bioclimatic design. Plants can mitigate the effect of urban heat islands and may also reduce the energy demand of buildings. A new trend in urban greenery is the creation of urban farms — using green areas in buildings to grow crops and vegetables.

In terms of façade applications, green walls are becoming more popular as a design feature. In fact, plant cover within a building has long been used for its decorative and thermal effects. Sky gardens can reduce thermal load on the occupied space below. These also provide amenity and create a cool semi-outdoor space for building occupants. Semi-outdoor spaces with vertical green screens can also provide shading from the sun and contribute to the garden ambience. Also, transpiration by plants extracts heat from the surrounding air and lowers the surrounding air temperature. Finally, the inclusion of exterior climbing plants can lead to a noticeable decrease in chemical contaminants and air pollutants.

Advantages of green walls:

- Cools building significantly, because shading and evapotranspiration remove heat.
- Visually attractive.
- Provides environmental benefits — stormwater attenuation, reduces urban heat island effect, improves air quality, and insulates the building.
- Protects the wall of the building from heavy rain.

3.2. Green buildings

3.2.1. *National library*

One of the building to achieve the highest Green Mark Platinum accolade is none other than the National Library Building. The building also won multiple other awards for its environment-friendly features.

Fig. 9. Greenery-enhanced National Library.

Internationally recognised as a 'green' building, its deployment of various innovative 'green' features helps to keep the building operating in an energy-efficient way and to do its part for a more sustainable environment. Some key features include the bioclimatic vegetation and landscaping to improve the indoor thermal environment (Fig. 9) and a lighting control system that controls the lighting according to surround illumination. (i.e. switches off lighting when there is sufficient natural lighting). In addition, the building is also greatly shaded to reduce solar heat gain through façade.

Key green features

- Building is oriented away from the East–West sun, combined with sunshading features on the West face of the building as an additional shield against solar heat gain and glare. Sunshading features include low-emissive double-glazed glass panel facade and large overhangs on the external facade.
- Light shelves that extend into the library space reflect sunlight further into the building.
  This optimises daylight and thus reduces the use of artificial lighting.
- Daylight sensor and automatic binds are installed and used together at the building facades.
- Motion sensors and energy-efficient lightings are installed in public toilets. They will be switched on only when required after closure of the library.

- Rain sensors are used as part of the automatic irrigation system for landscaped areas. Water-efficient taps and cisterns are also used to conserve water.
- Extensive landscaping, sky terraces, and roof gardens are utilised to reduce local ambient temperature.
- Open Plaza area located at the first storey allows natural ventilation and daylighting. Stack effect is observed at the open Plaza (i.e. the air from the sides of the building is drawn upward between the link bridges and in between the two blocks).
- Energy monitoring is via BMS (Building Management System).

3.2.2. HDB flats go ecofriendly: Punggol Eco-Town

The new Punggol Eco-Town is the first eco-development in the city-state, and it is a showcase for urban living solutions by the Housing & Development Board (HDB). As a testbed for technologies, it fosters the ecofriendly living and integrated communities. The award-winning Treelodge apartments utilise sustainable energy and smart design to citizens' life.

> "Instead of providing a functional and utilitarian kind of infrastructure we designed something that is beneficial and useful for the public to enjoy. We have incorporated various new technologies and solutions to encourage a green lifestyle for residents. HDB hopes to be a sustainable development solution hub so that we can share similar experience and knowledge with other cities of high-rise and high-density built environment."
>
> Mr Alan Tan Hock Seng,
> Director, Environmental Sustainability Research, HDB (2013)

Treelodge @ Punggol is Green Home where residents can live in an environment-friendly atmosphere. It introduces environmental features that embrace Singapore's hot climate conditions by employing both passive design strategies and green building technologies to achieve efficiency in energy, water, and waste management.

Key features

(a) Passive Design Strategy

Treelodge @ Punggol faces north-east and south-west wind directions. All the buildings are strategically orientated to face the prevailing winds to maximise natural ventilation. Windows are also designed to maximise natural lighting and minimise solar radiation from the sun into the units from the east-west directions.

(b) Greenery

The provision of greenery relieves urban heat island effect, thus reducing cooling load and energy consumption.

- Green roof is provided in ecoprecinct which reduces heat absorption on roof surfaces that enhances the comfort of residents living in the unit.

- Vertical greening and façade greening are other strategies introduced to cool off the block.

(c) Energy Management

Treelodge @ Punggol adopted energy-efficient technologies to optimise energy usage and reduce maintenance cost. This not only optimises energy usage but also explores the use of renewable energy source, reducing reliance on fossil fuels.

- Renewable energy technologies such as the solar panels are installed at the roof of the estate to generate energy to meet the demand of common area services such as lifts and lightings in the corridors and Eco-Deck.
- Motion sensors are also installed to allow energy-efficient lighting.
- "Cool walls" are incorporated to the block. The walls are enhanced with thermal insulation, with walls facing east or west direction to reduce heat radiation and improve the thermal comfort in the interior.

3.2.3. BCA Zero-Energy Building (ZEB)

Following the rise of green buildings, Zero-Energy Building (ZEB) has become the next challenge for engineers. As the name suggests, ZEBs are buildings with zero net energy consumption, meaning the total amount of energy used by the building is roughly equal to the amount of renewable energy created at site. ZEBs minimise the reliance on fossil fuel to generate energy and therefore reduces carbon emission.

BCA Academy takes the lead in becoming a three-storey workshop building that is retrofitted to be a ZEB. It also serves as a platform to testbed for green buildings in Singapore.

The building is retrofitted with photovoltaic technology which uses solar energy to power the building fully (Fig. 10). However, the building is orientated in east-west direction, where reducing solar heat gain is required. To overcome this, architects and engineers designed the sunshading system to minimise heat due to its orientation. Within the building are natural ventilation and lighting which minimise energy consumption.

Key features

(a) Energy-Efficient Envelop

Unlike normal clear glass, the building uses low-emissivity coating glass which increases energy efficiency of windows by decreasing the solar radiation transfer through the glass. Sunshading devices are strategically placed to significantly reduce the solar heat gain and improve the quality of natural lighting within the building.

(b) Lighting System

Energy-efficient lamp, automatic switching, and daylighting used to reduce the overall energy consumption due to artificial lighting.

Fig. 10. The solar PV cells on the roof of the ZEB.

(c) Active Control and Management

Building management system is used to control, monitor, and manage all the equipments installed in the building. With close monitoring of usage and occupancy patterns, energy use can be optimised while maintaining comfort and functionality.

(d) Air-Conditioning System

With technologically advanced chillers, variable speed drives, and personalised ventilation system in place, about 40% of energy is reduced for air-conditioning of the building.

(e) Fully Powered by Sun

Photovoltaic systems are installed to harness solar energy to generate electricity, hence fully power all the appliances and lighting in the building.

3.2.4. *Oceanfront @ Sentosa Cove*

The Oceanfront at Sentosa Cove is located within the vicinity of the Sentosa Island. The Oceanfront which resembles a shimmering sculpture will be the icon to captivate seafarers as they sail along the Singapore Straits to Sentosa Cove.

Apart from its aesthetics, various green designs engineered and installed made Oceanfront at Sentosa Cove a sustainable living area that was awarded the highest accolade, Green Mark Platinum.

For instance, computer simulations of Sun Path, solar insulation, and daylighting studies were used to determine the effectiveness of the building and interior layouts. Further consideration placed on the building orientation was employed to further reduce minimum west-facing. Energy-efficient air-conditioning system mounted with an inverter system was also installed. The inverter is engineered to continuously adjust its cooling and heating output to suit the temperature in the room. The inverter shortens system start-up time enabling the required room temperature to be reached more quickly. As soon as that temperature is reached, the inverter ensures that it is constantly maintained. With such efficient adjustment, a 30% reduction in energy consumption compared to a traditional on/off system is achieved.

In addition to the aforementioned features, Oceanfront at Sentosa Cove recycles water from apartment showers, baths, and washing basins for clubhouse toilet and landscape watering. The extraction of heat from air-conditioning is used in condensers for generating hot water for use in the clubhouse. Water-efficient features including rainwater harvesting system were employed for the irrigation of landscape. A twin-chute pneumatic waste conveyance system was engineered to allow separation of domestic waste and recyclable items, meeting sustainable practices and internal hotel recycling scheme. Finally, the use of prefabricated bathroom units was designed to reduce construction waste.

3.2.5. *Siloso Beach Resort, Sentosa*

The resort aims to set the environmental benchmark, providing visitors' experience a quality ecofriendly environment that will encourage them to adopt environmental activities, so that we can all live in a sustainable world.

Through adequate planning and sustainable engineering, hotel rooms located within the resort have taken extra care to keep the natural environment intact. For instance, buildings constructed were made to give way to trees.

Apart from the 200 and more original trees that were preserved on site, an additional 450 trees were planted to beautify the landscape. This aids by offsetting any previous damage to the land and to reduce the carbon footprint.

Various approaches employed to minimise the carbon footprint and sequester more carbon include greening the landscape through the planting of trees and other flora all around the resort. Rooms use compact fluorescent and LED lights, along with a key card system to activate the electricity.

Additionally, the resort aims to function as a testbed for advanced green technology. This comprises the employment of energy-efficient third generation modular heat exchange chillers system, which collects the heat dispersed from the air-conditioning process and employs it for heating water for rooms. In the kitchen, pilot trials are carried out to test a special energy-efficient water heater and a machine that uses bacteria to turn our food waste into fertilised water.

Besides, the hotel entails the concept of reduction in waste, wherever they find it. Old railway sleepers for stairs, reclaimed wood for repairing and making furniture were used.

3.2.6. Dawson Estate

Commissioned by the Housing & Development Board as an exploration of the future of affordable public housing, WOHA's (a Singapore-based architecture practice) public housing design for Skyville @ Dawson consists of 960 homes in Singapore. The project has been completed since early 2015 and its design focuses on three themes — community, variety and sustainability.

For the aspect of sustainability, the building was engineered and constructed using passive means. It avoids the use of energy-intensive solutions while employing the use of more advanced technology. Every unit is fully and naturally ventilated, with every room (including bathrooms and kitchens) having windows. Common areas, lift lobbies, and access walkways, are all naturally ventilated and lit. The apartments are cross-ventilated. Passive means for comfort are engaged for each and every individual unit. The walls of each apartment have vertical and horizontal sun breakers to shade both the walls and the windows; all windows have overhangs and special mid-height top-hung panels that direct breeze to seating height and allow the windows to remain open during the monsoon period. In addition, units were set to face north and south and have openings on all sides.

Furthermore, five different window types are used in the entire development. These designs create variety through the rearrangement of the modules, through color, light and shade. The site coverage is low, enabling a park to be created around conserved huge existing rain trees.

A 150-m long landscaped swale treats all the water before discharge and infiltration. Over 1.5 hectares of public gardens are provided. The roof landscape, vertical creepers to the car park, and sky gardens provide 100% green plot ratio.

PV panels are provided on the rooftop pavilions, enough to power the common area lighting. Dual refuse chutes for separation of organic and recyclable waste are provided at every apartment block. The design allows tropical living without air-conditioning, and all areas, whether common or private, are naturally ventilated and lit. The design is north–south facing and fully shaded. All units are cross-ventilated.

In addition, the design is fully precast. Rather than luxury of materials, the precast, painted design proposes resolution of social, technical, and aesthetic objectives as the creation of most value. Such construction is employed to avoid on-site waste and make construction efficient. The design uses quality of form, space, light, ventilation, and proportion for its impact. The block is perforated, folded, and studded with gardens to avoid the appearance of a large mass.

References

Building and Construction Authority. (2010). *Green Building Platinum Series: Building Planning and Massing.* Singapore: Building and Construction Authority.

Building and Construction Authority. (2010). *The Island's First Retrofitted Zero Energy Building.* Retrieved 2016, from https://www.bca.gov.sg/zeb/

Building and Construction Authority. (2011). *Build Smart: A Construction Productivity Magazine* (Vol. 4). Singapore: Building and Construction Authority.

Building and Construction Authority. (2013, August 1). *International Experts: More Benefits can be Reaped From Building Information Modelling*. Retrieved from BCA Newsroom: https://www.bca.gov.sg/Newsroom/pr01082013_IPE.html

Building and Construction Authority. (n.d.). *Singapore: Leading the Way for Green Buildings in the Tropics*. Retrieved 2016, from https://www.bca.gov.sg/greenmark/others/sg_green_buildings_tropics.pdf

Chin, D. (2013, May 18). *New Panel to Help Construction Sector Tackle Labour Crunch*. Retrieved June 29, 2016, from asiaone: http://www.asiaone.com/business/new-panel-help-construction-sector-tackle-labour-crunch

Furuto, A. (2012, March 13). *SkyVille @ Dawson/WOHA*. Retrieved 2015, from ArchDaily: http://www.archdaily.com/215386/skyville-dawson-woha/

Housing & Development Board. (n.d.). *TOWARDS A BARRIER-FREE ENVIRONMENT IN PUBLIC HOUSING*. Retrieved from https://www.bca.gov.sg/BarrierFree/others/HDB.pdf

Housing & Development Board. (n.d.). *Treelodge @ Punggol: Green Home, Healthy Living, An Eco-lifestyle Experience*. Retrieved April 16, 2017, from http://www10.hdb.gov.sg/ebook/treelodge/punggol.html

Inter-Ministerial Committee on Sustainable Development. (2009). *A Lively and Liveable Singapore. Strategies for Sustainable Growth*. Singapore: Ministry of the Environment and Water Resources/Ministry of National Development.

National Library Board Singapore. (2017). *National Library Building*. Retrieved March 5, 2017, from National Library Board Singapore: http://www.nlb.gov.sg/VisitUs/NationalLibraryBuilding.aspx

Palanichamy, M., Muthuramu, K., & Jeyakumar, G. (2002). Prefabrication techniques for residential building. *OUR WORLD IN CONCRETE & STRUCTURES*. Singapore: CI-Premier PTE LTD. Retrieved from http://www.cipremier.com/e107_files/downloads/Papers/100/27/100027054.pdf

Siloso Beach Resort, Sentosa. (2010). *Our Green Initiatives*. Retrieved 2015, from Siloso Beach Resort, Sentosa: Blending Life and Leisure with Nature: http://silosobeachresort.com/slideshow.aspx

Tung, S., & Choo, F. (2014, Jun 12). *Five Vertical Gardens in Singapore That Have Hit a High Mark*. Retrieved 2015, from The Straits Times: http://www.straitstimes.com/singapore/five-vertical-gardens-in-singapore-that-have-hit-a-high-mark

Chapter 6

Aerospace

1. Introduction

Today, Singapore Aviation comprises myriad success stories spanning all sectors of industry. We have a vibrant industry in which people grow, enterprises thrive, and ideas flourish, making Singapore an aviation hub of choice.

In the 50-odd years since Singapore's modern founding, we have made leaps and bounds in the development of our nation as an aerospace hub. More than 58 million passengers — more than 10 times our population — pass through Changi Airport in a year, and, on average, a flight lands or departs just under every two minutes (2016 figures). Similarly, Singapore is also a prime destination for cargo shipment via air, from fresh flowers to high-value goods. Since 2012, Changi Airport has also handled more than 1.8 million tonnes of air freight every year.

Singapore has developed into a nation with a comprehensive and efficient aerospace engineering sector. It is a S$8.3 billion aerospace engineering hub today (2014). Singapore offers a comprehensive range of maintenance, repair and overhaul (MRO) services and advanced manufacturing capabilities, making us a leading aerospace hub in the Asia-Pacific. This is evidenced by being home to more than 100 world-renowned aerospace companies and related engineering equipment manufacturers. The growth of the sector has been supported by the creation of a world-class business infrastructure and the synergies from cluster integration, such as economies of scale, collaboration with leading players and a highly-skilled talent pool. Singapore has now earned a good reputation as a result of its world-class aircraft support services, excellent Maintenance, Repair and Overhaul (MRO) capabilities, investment in maintenance and repair training, as well as in aerospace education, research and development, and establishing comprehensive cargo logistics capabilities.

The changes that have shaped the wings of aviation in Singapore in the last 50 years, especially the last 20 years, is the result of meticulous planning and intuitive foresight from the planners and policymakers, as well as hard work and contributions from the engineering community. Some of the key success factors contributing to this are: A capable government and civil service that created a stable political and business landscape, a global outlook, liberal economic policies that extended into the aviation realm, a safe and secure environment underpinned by the rule of law, a motivated and skilled workforce, a capable engineering personal, and a keen focus on quality.

With Singapore strategically positioned on historical trade routes that traversed Europe, the Middle East and India, crossing to the Far East and extended south to Australia and New Zealand, Singapore is well-placed as a regional commerce and logistics hub from which global companies operate and tap the dynamic growth in the Asia-Pacific.

Over the last 50 years, we saw the multinational companies that first invested in Singapore setting the stage for further economic growth, spurring a rise in air passenger travel and cargo volumes through Singapore and attracting even greater investments. This growth in turn kick-started the provision of engineering-related services that, although borne out of necessity to service Singapore's burgeoning airline and air force, were quick to adapt to the demands of a rapidly growing international air carrier industry and later becoming full-fledged global providers of aerospace-engineering services.

2. Changi Airport

Growth in demand for air services drove rapid airport development that saw successive innovations in terms of infrastructure, facilities and services, in which established Changi Airport as a global brand and benchmark of quality. Great efforts also have been focused on developing highly skilled human capital to enable future growth through the support of a variety of educational options. These innovations have established Singapore Airlines and aerospace-engineering companies like ST Aerospace, acknowledged as leaders in their respective industries.

Changi Airport is an icon for service excellence, synonymous with good customer service delivery and strong capabilities to create exceptional journeys that deliver a memorable airport experience for airport users. Opening its doors in July 1981, the airport continues to be lauded for its unyielding commitment to innovation, excellence and its desire to constantly raise the bar. Changi Airport remains the world's most awarded airport, collecting another 25 *Best Airport* awards in 2010. To date, Changi is the proud recipient of more than 370 awards since its opening 30 years ago.

Over the past three decades, Changi Airport has also seen its traffic increase fivefold. The Airport has continued to adopt a proactive policy of building in anticipation of future traffic demand and providing ample capacity to service even more passengers and airlines. Its facilities have been constantly upgraded so as to maintain its competitive edge as an attractive airport for travellers and as an efficient operating environment for airlines.

From a single terminal in its formative years, Changi Airport now operates four terminals — Terminals 1, 2 and 3, as well as the Budget Terminal in year 2014. From 2017–18 onward, Budget Terminal will be reconstructed and renamed as T4.

2.1. *Changi Airport's significant milestones*

1981	Officially declared opened
1986	Surpassed the 10 million passenger milestone
1990	Terminal 2 opened for flight operations on 22 November. Skytrain introduced to provide a swift link between Terminals 1 and 2
1994	Surpassed the 20 million passenger milestone
2002	Changi Airport Mass Rapid Transit (MRT) station officially opened on 27 February, improving accessibility to the airport
2004	Crossed the 30 million passenger mark
2006	The first Budget Terminal for low cost carriers in Asia opened at Changi
2007	World's first A380 passenger flight took off from Changi Airport
2008	Terminal 3 commenced scheduled flight operations on 9 January
2009	Changi Airport corporatised. Changi Airport Group formed to undertake key functions like airport operations and management and air hub development
2010	Named *Best Airport In The World* for the 23rd year running by Business Traveller (UK). Passenger movements breached the 42-million mark, a record for annual traffic
2011	Connectivity to a sixth continent established with the launch of flights by Singapore Airlines to Sao Paulo, Brazil

2.2. *Historical development of Changi Airport*

During wartime in the mid-1940s, the Japanese occupation force had constructed rudimentary landing strips near Paya Lebar. This was a key development that would later influence the siting of Singapore's present day civilian airport. However, poor soil conditions made construction of proper runways a difficult proposition at the time.

Paya Lebar Airport represented the most modern airport of its day and was easily the best such facility in Southeast Asia. It also provided an important learning experience for Singapore — it was here that the soon-to-be independent nation experienced the great changes to the international flow of people and goods that the dawning 'jet-era' spawned.

During this time, air cargo was also a burgeoning business. Larger, wide-bodied planes made it feasible to carry larger amounts of cargo in addition to passengers. This resulted in greater demands being placed on cargo handlers, fostering the development of a more efficient and professional industry.

Recognition of the importance of cargo for business inspired Singapore Airlines to build dedicated warehousing and handling facilities at Paya Lebar Airport. Airport planners also correctly anticipated the rapid growth of airfreight through Singapore and the importance it would play in attracting investments from multinational corporations to the young nation, and they stepped up to cater to these emerging demands.

In the short time after the war, passenger numbers continued to climb at a rapid rate, both globally and in Singapore, as new jet aircraft — including the revolutionary Boeing 747 jumbo jet — brought the global movement of people into the mainstream. By 1960, Paya Lebar was handling more than 300,000 passengers and 30,000 aircraft movements annually and enjoying unprecedented connectivity to major airports around the world.

Since independence, our national carrier, Singapore Airlines, SIA, was established, and was fast moving to tap the brisk growth in air traffic. Also, international airlines began calling on Singapore in substantial numbers. By 1970, the number of passengers swelled to 1.7 million and, by 1975, it had tipped the four million mark. Once again, pressure mounted on the existing aviation infrastructure.

In late '70s, a decision was initially made to add a new runway and further expand the terminal building at Paya Lebar. However, the plan was ultimately scrapped and authorities voiced the history-making decision to create a brand-new civilian airport at Changi. This was enabled by engineering advances achieved since the last inspection of the land there decades earlier. Since then, land reclamation works was carried out at Changi to make room for the new airport. Extensive ground treatment works were researched and successfully carried out as the newly reclaimed land is underlain with soft seabed clay which will undergo excessive amount of continuous settlement for a long time, if not properly treated. A new ground treatment method called Vertical Drain methods was employed for the first time in Southeast Asia to treat this soft clay. The reclaimed sand obtained from the nearby Bedok hillcut and the seabed dredged sand was also have to be treated by another new method — Dynamic Compaction method. The air traffic control tower was designed and constructed in an innovative manner, the 16-sided, 78-metre tall structure remaining an iconic symbol of the airport to this day (Fig. 1).

Since its official opening days, Changi Airport continues to expand and upgrade. By 1990s, T2 and T3 were constructed and linked to T1 via another engineering achievement, the Skytrain.

2.3. *Future development at Changi Airport*

In the pipeline are three highly anticipated projects: Terminal 4 and Project Jewel, as well as the Changi East Development which includes a third runway and a fifth terminal for Changi Airport. Terminal 4 (Fig. 2), which will be operational towards the end of 2017, will roll out an extensive suite of FAST (fast and seamless travel) initiatives to automate airport processes and improve productivity. These include self-service and automation options, such as self-check-in, self-bag drop, automated immigration clearance and self-boarding at the departure gates. In addition to Terminal 4, work also begun on Project Jewel — an iconic 3.5-hectare mixed-use complex at Terminal 1 — which targeted to open by 2018. Project Jewel will augment Changi Airport's offerings and strengthen its position as an international air

Fig. 1.   The iconic Changi Airport control tower.

hub with a wider range of travel-related services and facilities for airport operations, retail and leisure attractions (Fig. 2).

To be ready by the late-2020s, the 1,080-hectare site in Changi East (Fig. 3) is slated to be a major and highly complex development that will increase the airport's capacity to 135-million passenger movements per annum (mppa) and preserve its position as an air hub.

3. Air Cargo Service

Air cargo has been an integral part of the aviation landscape for over 80 years. In 1930, Seletar Airbase was opened to commercial aircraft. It was there that, on 11 February that year, a Dutch-made, tri-engine monocraft carrying eight passengers and a cargo of fresh fruits, flowers and mail landed. The historic flight marked the beginning of commercial civil aviation here and the beginning of the airfreight business. By 2015, Singapore boasted an annual air cargo throughput of over 1.9 million tonnes per year, making Changi Airport the seventh busiest cargo handling airport in the world. A number of key developments gave the air cargo industry the real impetus for growth to become what it is today, including more

Fig. 2. Artist's impression of the Terminal 4 departure area (top) and Project Jewel (bottom).

efficient and cost-effective movement of goods due to larger, wide-body aircraft. The growth of manufacturing in Singapore was another contributing factor.

It began in the early to mid-1960s when prominent American company Caterpillar began bringing equipment and spare parts into Singapore and storing them here for its tractors and other products which set the scene for the establishment of the logistics of air cargo distribution and cargo hubbing, long before the concept was truly realised.

Fig. 3. Terminal 5 (centre-right, beside the two-runway Changi Airport complex) is scheduled for completion in the late 2020s.

Cargo ground handling also began its ascendancy on the island with the formation of SATS Ltd (SATS). In the early years, ground-handling services were provided by a division of Malayan Airways, which became Malaysia-Singapore Airlines (MSA) in 1967. Five years later, MSA ceased operations and paved the way for two new entities — present-day Singapore Airlines (SIA) and Malaysian Airlines.

As SIA concentrated on its core business of running an airline, SATS evolved naturally as a separate, yet wholly owned, subsidiary. In 1977, SATS opened its first airfreight terminal at Paya Lebar Airport capable of handling 160,000 tonnes of cargo a year.

The key impetus for the development of the air cargo industry in Singapore was the push by then-Prime Minister Lee Kuan Yew to attract foreign direct investments by multinational corporations (MNCs). The move provided a powerful boost to the air cargo sector which grew in tandem with the rapid growth of the Singapore economy, fuelled by the MNCs' economic contribution. Today, over 7,000 MNCs continue to provide a powerful economic engine in Singapore.

The 1960s also marked the start of Singapore's freight forwarding sector and saw the development of a handful of companies — almost all foreign — that performed the crucial tasks of facilitating the movement of cargo from the exporter or shipper to the airline and the reverse procedure to the buyer at the other end. These forwarding companies also played a crucial role in training and imparting the knowledge and skills of the nascent forwarding industry to Singaporeans, many of whom went on to form their own highly successful freight forwarding outfits.

By the 1970s, freight forwarding companies numbered around 70. As the air cargo industry was on the verge of taking off in Singapore, Mary Wu, Managing Director of Singapore Baggage Transport Agency Pte Ltd, saw the need to represent the interests of Singapore cargo agents in an organised manner. She gathered the resources of the six pioneering cargo agents to establish the Singapore Aircargo Agents Association (SAAA) in 1971. It helped local agents adjust and upgrade their services to meet the demands of the changing business environment, particularly with competition from new international agents setting up their offices in Singapore and logistical problems such as in the declaration of trade and customs documents. Today, the number of freight forwarders here has leapfrogged to over 400 and consists of both foreign multinational forwarding companies and Singapore-based businesses.

When Changi Airport opened in 1981, the air cargo industry was on the cusp of a new era. To consolidate and take advantage of what was now clear to policymakers as very favourable conditions for a burgeoning industry, the government set about building the necessary infrastructural facilities at Changi Airport, as well as regulatory procedures that would ease the flow of business operations. Among the new infrastructure was the Free Trade Zone (FTZ) within the Cargo Complex, where documentation and red tape were reduced to a minimum. New cargo warehouses were built with sufficient capacity and latest cargo handling equipment for speedy and efficient handling. Forwarding agents were also housed inside the FTZ to facilitate handling and sorting of cargo, boosting the creation of a professional forwarding industry.

Cargo ground handling was similarly set on a path of greater efficiency and growth with new facilities. SATS made the move to Changi Airport after investing S$147 million in facilities that included two new, modern airfreight terminals.

It has also been a positive growth story for Changi International Airport Services (CIAS), which undertook the provision of cargo, security, passenger, baggage, ramp and technical ramp services at Paya Lebar Airport from its inception in 1978, ultimately growing to its current 1,500-strong workforce.

Cargo continued its rapid growth alongside the young Singapore economy and to better tap the growing market, SIA created a separate cargo division to complement its passenger-carrying business in 1992. From this point forward, it was a strategy of continually adding and upgrading infrastructure and services at and around the airport. The government's pro-business outlook and sustained moves to encourage global companies to set up regional bases here continued to drive air cargo, just as air cargo made it possible for these companies to operate and even thrive in Singapore.

From simple airfreight being manually hauled from warehouse to awaiting aircraft in the early days to Changi's focus on efficient air cargo handling, the industry became more interlinked and multi-modal as it helped transform Singapore, following global trends, into a key global supply chain and logistics hub. Robust infrastructure have contributed to the establishment of many manufacturing and regional distribution centres here that rely on fast and efficient movement of cargo, including consumer goods, electronics, chemicals and life sciences companies.

Singapore has been investing in a robust infrastructure of cargo facilities, as well as working closely with airport agencies and the cargo community to allow for speedy, efficient and hassle-free transfer and clearance of goods. The 47-hectare Changi Airfreight Centre (CAC) is a security-restricted area at Changi Airport that includes a 24-hour Free Trade Zone where transhipment cargo can be broken down and reconsolidated with minimal Customs formalities. There are eight airfreight terminals at the CAC, with two dedicated Express and Courier Centres to accommodate time-definite shipments.

A key development was the establishment of the 26-hectare Airport Logistics Park of Singapore (ALPS) within the Free Trade Zone (FTZ), which allows third-party logistics providers to leverage Changi Airport's excellent connectivity and superior handling efficiency for quick turnaround of value-added logistics and regional distribution activities.

More recent developments have seen the addition of a cool-chain perishables facility in mid-2010 that caters to the growing and highly specialised temperature-controlled cargo segment, which includes food, pharmaceuticals and bioscience products. Built and operated by SATS, the S$12 million Coolport@Changi is Singapore's first on-airport perishables handling centre and has an annual operating capacity of around 250,000 tonnes.

4. Singapore Airlines

The importance of focusing on international passengers and traffic from the very earliest days was adopted by the architects developing Singapore as a global aviation hub. Recognising that to remain as a strategic aviation hub in the face of the rapidly changing global aviation environment, Singapore understands well the need to find ways to ensure that civil aviation and its assorted capabilities, continue to grow and add value, not just at home, but in the region and across the globe.

Airlines are the bedrock of the aviation industry. International connectivity has always been a priority for Singapore for social and economic reasons. The country's first international air route was established in 1930 when Seletar Airport began hosting scheduled commercial services between London and Batavia (present-day Jakarta) that was operated by Koninklijke Nederlandsch-lndische Luchtvaart Maatschappij (KNILM, also known as Royal Netherlands Indian Airways), an airline belonging to the Dutch East Indies.

Following that inaugural air link, air traffic through Singapore began growing at a steady pace, necessitating a new airport, which was constructed at Kallang and started operations in 1937. Kallang Airport is perhaps best remembered as the airport of the colonial 'glory days', which saw spacious flying boats landing on the Kallang River and discharging their well-heeled passengers at the grand terminal.

Singapore got its start in scheduled airline service courtesy of Imperial Airways, the flagship airline of the British empire at the time, when it started a regular service between London and Darwin via Cairo, Karachi, Calcutta, Singapore and Jakarta. It gave an early clue to Singapore's rise to importance as an air transit hub. As passenger traffic began to build, another scheduled airline — Wearnes Air Services (WAS) — began operations between Singapore and towns in the Malay Peninsula. The establishment of WAS showed the early potential of Singapore not only as a transit stop for international carriers, but also as a hub for regional flights. Both aspects continue to play an important role in Singapore's position as an air hub today.

However, demand soon outstripped Kallang Airport's capacity in the 1940s. Larger, faster and heavier aircraft carrying ever-greater numbers of passengers became the norm, due to a combination of a post-war economic boom and concomitant surge in air traffic and new aircraft technology. Lack of sufficient infrastructure — in terms of passenger terminal and runway capacity — to cope with the surge meant that the hunt was on again for a new airport location.

With the outbreak of World War II in 1939, commercial aviation came to a halt for over half a decade. However, its resumption heralded a new era of voracious growth for the global aviation industry and, along with it, for flights through Singapore. Of great significance to Singapore was the establishment of a new home-grown airline, the Malayan Airways Limited (MAL), with ownership by the Ocean Steamship Co., Straits Steamship Co. and Imperial Airways.

It may seem odd for maritime companies to be involved in starting up an airline, but so clear was the potential of air travel at the time that it threatened to eclipse ships as the mode of intercontinental travel. MAL started operations in May 1947 and no one could have known then that it was to be the precursor to one of the top airlines in the world — Singapore Airlines (SIA).

In 1963, the Federation of Malaysia was formed and Malayan Airways was renamed Malaysian Airways. Thereafter, it was renamed Malaysia-Singapore Airlines or MSA, complete with a notable fluorescent yellow livery. In a move that seemed to foretell the future, MSA's primary hub was established at Paya Lebar Airport, taking advantage of its sophisticated facilities and support systems. The carrier's fleet, infrastructure and corporate development continued rapidly with MSA's operations stretching to Australia, India and Europe.

The 1960s witnessed a period of rapid growth of the air travel industry, due in a large part to the emergence of jet-powered aircraft that could carry larger numbers of passengers for greater distances, effectively bringing global air travel to the masses. In what was to be a lasting legacy for the precursor of SIA, MSA anticipated these changing trends and became an early operator of Boeing's new B707 jet aircraft, which is credited with facilitating this new era of mass travel.

Following the unsuccessful merger of Singapore and Malaysia, the two nations began charting very different courses for their respective futures and their vastly different national priorities were reflected in their ideas for development of their joint airline. While Malaysia was intent on focusing on domestic services, Singapore was keener on developing international routes. This philosophy underlines the strategy for the development of aviation in Singapore.

MSA was thus restructured into two separate national carriers: Malaysian Airline System (MAS) for Malaysia and Singapore Airlines (SIA) for Singapore. Both airlines started operations on 1 October 1972.

SIA inherited a penchant for expansion and foresightedness, which has become the airline's hallmark. This was clearly demonstrated by the airline's bold order for four Boeing 747-200s, which had only debuted two years earlier, making it Southeast Asia's first 'Jumbo Jet' operator.

Affirming SIA's decision to invest in the largest commercial aircraft at that time, Singapore's passenger numbers rose from 1.7 million to four million between 1970 and 1975. The jump in traffic ultimately prompted the construction of a brand-new airport — Changi International Airport — which opened in 1981. The seed of SIA's current success is evident in its foresight to break flight service "barriers", becoming the first carrier to offer free drinks and complimentary headsets onboard, among many others.

SIA even became a brief operator of the revolutionary Concorde supersonic jet in partnership with British Airways between 1977 and 1980, mainly on the London-Bahrain-Singapore route. Although the service was short-lived, the endeavour was nothing less than a marketing coup for SIA as it was one of only four airlines in the world to have operated the futuristic, premium aircraft. This once again emphasised the carrier's unique appreciation of raising inflight service standards in the air travel market.

Scheduled regional services were introduced in February 1989 with the establishment of Tradewinds Airlines. Flights were available to six destinations in the region from Changi Airport and one from Seletar Airport to Tioman in Malaysia. With growing regional business, the carrier was re-branded SilkAir in 1992. The creation of Tradewinds and SilkAir fulfilled demand from the regional leisure travel sector. While regional travel has now become the principal domain for low-cost or budget carriers, SilkAir still thrives with its provision of full-service regional travel.

With cargo flowing through Singapore and the Asian region growing in leaps and bounds, SIA created a separate cargo division to complement its passenger-carrying business. The separate subsidiary became responsible for all of the carrier's cargo activities involving both freighters and cargo carried in the belly-hold of its passenger fleet. SIA Cargo quickly rose in the prominence in the industry, and now ranks as the world's ninth largest player in terms of cargo carried (scheduled freight tonne-kilometres).

Living up to its increasingly well-known reputation, SIA placed an early order for Airbus' new 'super jumbo' double-deck A380. The carrier became the first in the world to fly this jet commercially in October 2007.

A recent major airline development for Singapore and the region was the emergence of the low-cost carrier model. This has long been in place in Europe and the United States but it was delayed in Asia until sufficient liberalisation unlocked the door for low-cost carriers.

In 2004, Valuair launched its maiden flight and became the catalyst for the formation of more Singapore-based low-cost airlines. In rapid succession, Tiger Airways and Jetstar Asia Airways also quickly established their presence at Changi Airport.

5. Industry Development

5.1. *Maintenance, Repair and Overhaul (MRO) services*

In 1937, aircraft maintenance was performed by engineers and mechanics at Kallang Airport. From those early days at Kallang Airport where a handful of Singaporean engineers and technicians performed routine maintenance on aircraft, the industry has grown in sophistication, requiring a highly educated and skilled workforce, employing over 20,000 people, making it a significant economic contributor.

MRO success is intricately intertwined with other parts of Singapore's aviation industry. As Singapore grew in prominence as an international air crossroads, the MRO sector grew in step — not only in terms of volume, but in depth and breadth as well. Singapore is now Asia's most comprehensive MRO hub, commanding a 25 percent share of the Asia-Pacific MRO market and a compounded annual growth rate of 8.6 per cent (2016).

The success of this industry has its roots in both historical developments and foresight by government economic planners, which led to concerted efforts in creating an environment conducive for MRO growth here. This effort has focused on a multi-pronged approach for developing cornerstone homegrown players, attracting major global MRO players and developing small- and medium-sized local companies to support these larger entities.

With the establishment of Singapore Airlines in 1972 — and before that, its precursors Malayan Airways and Malaysia-Singapore Airlines — the young and growing carrier relied on a simple line station operation at Paya Lebar Airport. This quickly grew to a major maintenance and overhaul complex at the airport, forming the early beginnings of the industry in Singapore.

The 1960s and 1970s were an exciting era for aviation, both globally and in Singapore, as many technological advances came to fruition in the commercial aviation field, fostering rapid growth in commercial aircraft traffic. This saw commensurate growth in other areas of commercial aviation, such as airports and, more importantly, engineering and maintenance activities. Aside from growth in Singapore Airlines' facilities to accommodate the carrier's growing fleet of aircraft, a number of companies were also established to tap into aircraft sales and distribution, repair and overhaul of aircraft, components, instruments and associated parts for the increasing number of international carriers flying through Singapore.

In 1975, a milestone was reached with the formation of Singapore Aircraft Maintenance Company (SAMCO), which was created to support the new Republic of Singapore Air Force. This was to have a far-reaching impact on the development of this sector in Singapore as it formed the basis for a move into home-grown advanced aerospace engineering capabilities for commercial aircraft. The maturing of the various aviation MRO capabilities led to SAMCO's expansion, eventually becoming what is today known as Singapore Technologies Aerospace (ST Aerospace), a cornerstone of the local MRO industry.

Recognising the economic potential of offering aero engineering services to the growing number of air carriers passing through Singapore, the government began ramping up efforts to promote aerospace activities amongst domestic players and by attracting foreign manufacturers to set up maintenance outfits here. The first component repair facility started operations in 1978; SAMAERO Company Pte Ltd was also set up the same year.

By the 1990s, aerospace-engineering activities were in full swing. SIA's purchase of B747s and B777s during this period attracted major aircraft engine manufacturers such as Pratt & Whitney and Rolls-Royce to set up huge engine overhaul centres in Singapore. These in turn attracted businesses to set up ancillary services that catered to them.

The continued purchase of Rolls-Royce engines by SIA was a further catalyst for the company to set up a major engine assembly plant in Seletar Aerospace Park (SAP). These activities, combined with wise government policies, set the pace for the remarkable MRO growth in the 1990s and 2000s.

With the opportunities this period of rapid growth presented, SIA moved to hive off its engineering capabilities by creating a new subsidiary, SIA Engineering Company (SIAEC), in 1992.

With over 30 years of aviation experience, SIA Engineering Company (SIAEC) is one of Asia's leading MRO companies renowned for its total aircraft maintenance solutions. SIAEC specialises in servicing all the new generations of Boeing and Airbus aircraft fleets, including the most technologically advanced airframes such as the A380 and B787. To ensure a fleet's seamless operations, SIAEC provides a wide range of maintenance services including heavy maintenance checks, light and transit checks, fleet technical management, inventory pooling and repair management, cabin retrofits and aircraft modifications, and painting services, contributing substantially to Singapore's aerospace industry output.

Furthermore, SIA developed the Tailored Support Programme with Airbus in 2008. The first of its kind in the world, it was a customised maintenance and engineering support package for SIA's Airbus fleets. Subsequently, a similar programme between SIA and Boeing for the Boeing fleets further rooted the two world's biggest airframe manufacturers to Singapore as they set up Fleet Management Offices in Singapore to operate these programmes.

Another industry stalwart, ST Aerospace, the world's largest commercial airframe MRO provider, has embarked on several initiatives to expand and value-add to its capabilities. In 2014, it opened its new aviation centre at SAP, housing a comprehensive suite of air charter, ground training, flight training and support facilities. ST Aerospace is the only integrated aviation service provider in SAP.

Moving up the value chain, ST Aerospace provides complete turnkey solutions for VIP cabin interiors for Airbus Corporate Jets and Boeing Business Jets, as well as cabin interior reconfigurations for a wide variety of commercial aircraft. ST Aerospace has also recently designed its own aircraft seats.

Today, Singapore has created an effective aerospace ecosystem, with activities ranging from logistics to design, from MRO to manufacturing, from sub-components to airframes, and from transit certification to fleet management all under one roof.

5.2. *Homegrown aerospace companies and local R&D efforts*

This most recent decade has seen rapid development with both ST Aerospace and SIAEC expanding their global footprints. SIAEC also formed an important joint venture with Rolls-Royce in 2001 — the Singapore Aero Engine Services Ltd, which is the Rolls-Royce Centre of Excellence in the Asia-Pacific for the repair and overhaul of Rolls-Royce Trent engines.

Importantly, these two Singapore companies have acquired the full spectrum of airframe, components, avionics and engine maintenance and overhaul. This marked an important achievement for not just the companies, but Singapore as a whole because of the highly proprietary nature of the technical data needed for such work. The transfer of valuable intellectual property signalled that the Singapore industry had matured and reached a very high level of quality and safety.

The importance of Singapore as a global aerospace engineering hub was further reinforced by the addition of world-leading aero engineering (OEMs) original equipment manufacturers like Boeing, EADS (Airbus), Thales, Honeywell and Parker Aerospace. The industry was also increasingly bolstered by a large precision engineering supplier base of local small and medium enterprises (SMEs), which provided crucial support for aerospace manufacturing and MRO activities.

To further the growth of the industry, the government, through various agencies, embarked on an intensive plan to further develop aerospace engineering capabilities here. This includes ongoing efforts to attract investment from global aerospace majors, assist SMEs to compete through initiatives like the Capability Development Programme, promote business alliances through the iPartners Programme and encourage the adoption of quality management systems like the AS9100.

The government also furthered efforts to promote Singapore as an aerospace hub through event platforms such as the Singapore Airshow, as well as a focus on education in order to develop an engineering talent pool, which included a S$76 million initiative to reinforce the specialist area of precision engineering education.

Important steps to establish a research and development (R&D) cluster were also undertaken. One of the flagship efforts was the Aerospace Programme established by the Agency for Science, Technology and Research (A*STAR) in 2007.

The programme brings together global aerospace players, local industry partners and the A*STAR research institutes to work on pre-competitive aerospace research. Through collaboration in a consortium, the members were able to drive aerospace innovation and benefit from synergies in the research.

Through government support, innovation and the ability to leverage the country's excellent connectivity and success as an air hub, Singapore's aerospace industry has established itself as a respected global leader capable of providing nose-to-tail capabilities.

Today, the island's MRO cluster consists of over 100 companies and to further develop and consolidate this solid base of expertise and business, the government created the 320-hectare Seletar Aerospace Park (SAP) with the aim of setting the industry on course for its next generation of growth.

The integrated, 'plug-and-play' Seletar industrial space has been designed to cater to a wide range of aerospace activities — from aerospace MRO to the design and manufacture of aerospace systems and components to business and general aviation activities, such as a training campus for pilots, aviation professionals and technical personnel. It is currently home to more than 30 aerospace companies, including Airbus, Bell Helicopters, Execujet, Fokker Services Asia, Hawker Pacific Asia, Jet Aviation Singapore, Pratt & Whitney, Rolls-Royce, and ST Aerospace Engineering.

In 2003, the industry established the Association of Aerospace Industries (Singapore) (AAIS), dedicated to the promotion of competitiveness within Singapore's aerospace industries. As part of its aim of positioning Singapore as the aerospace hub of choice in the region, the AAIS now has resource capabilities to assist companies in need of certification, accreditation and career promotion and progression. The AAIS, in partnership with industry and government, undertook an initiative in 2009 to map the local aerospace industry landscape and determine its strengths and weaknesses, along with strategies for addressing them. The study's recommended efforts to enhance this sector lie in Singapore's strong precision engineering support cluster, which is well-equipped to support large OEMs. With the SAP coming online and a new roadmap, the local aerospace cluster looks set for more decades of growth and development.

Aerospace research and development (R&D) activities have also grown significantly over the past few years. With more programmes and incentives to spur innovation and growth, these have helped to further cement Singapore's role as an aerospace hub in the region.

The Economic Development Board (EDB) introduced the Industrial Postgraduate Program (IPP) to build up a pool of postgraduate manpower with critical R&D skill sets. Under the programme, PhD students spend the majority of their time working on research projects in participating companies. A company that has participated in the IPP is EADS Innovation Works. It established its first research and technology centre outside Europe in Singapore and undertakes R&D for EADS' businesses, which includes aerospace.

Additionally, Singapore has nurtured a robust intellectual property regime that is attuned to the needs of companies to protect invention and innovation. As such, it has successfully attracted aerospace companies and aviation-related R&D centres that harness Singapore's strong base of world-class R&D facilities, research institutes and universities.

The Rolls-Royce @ Nanyang Technological University (NTU) Corporate Lab (Fig. 4) that was launched in July 2013, is one such example. The lab helped Rolls-Royce source for new innovations in areas such as electrical power and control systems, manufacturing and repair technologies, and computational engineering.

The S$75-million partnership between the university, Rolls-Royce and the National Research Foundation, Prime Minister's Office, Singapore, has been in the process of jump-starting over 30 new projects up till 2018, tripling the existing number of projects between Rolls-Royce and NTU. This is in addition to Rolls-Royce's S$700-million Rolls-Royce Seletar Campus, which includes a state-of-the-art assembly and test facility for Trent aero engines, an advanced manufacturing facility

Fig. 4. (Left to right) NTU Chief of Staff Prof Lam Khin Yong; NTU Provost Prof Freddy Boey; then-NTU President Prof Bertil Andersson; then-Minister of State for Finance and Transport Josephine Teo; Rolls-Royce Director of Research and Technology Prof Ric Parker; National Research Foundation CEO Prof Low Teck Seng and Rolls-Royce Regional Director for ASEAN & Pacific Jonathan Asherson officially launching the Rolls-Royce @ NTU Corporate Lab back in 2013.

for hollow titanium Wide Chord Fan Blades, as well as cutting-edge research and training facilities.

With its strategic location, competitive tax rates, sound legal, workforce and economic infrastructure as well as liberal trade and aviation policies, Singapore offers aviation organisations an advantageous base and opportunity to spread their wings here. The country's ongoing efforts in strengthening its infrastructure to meet business and investment needs have also put it in good stead with aerospace companies.

Rolls-Royce has a long and successful history of working in Singapore and has maintained an office here since the 1950s. Today, Singapore is a key business hub for Rolls-Royce, and all four major business sectors — Civil and Defence Aerospace, Marine and Energy — are strongly represented. As of December 2013, the Group, along with its joint venture partners, accounts for over 15 percent of the country's aerospace output.

For Pratt & Whitney, a world leader in the design, manufacture and service of aircraft engines, space propulsion systems and industrial gas turbines, being closer to customers means extending its value chain in Singapore. With a comprehensive presence of having established seven facilities already in the Republic, the company is putting in nearly US$110 million to develop another two facilities — one for MRO and engineering, and the other for manufacturing — at Seletar Aerospace Park.

For Rolls-Royce, the group and its joint venture partners currently employ more than 2,200 people in Singapore. At Rolls-Royce's 154,000-square-meter Seletar Campus facility, nearly 1,000 people are employed across various functions, and over 90 percent are Singaporeans or PRs.

For Pratt & Whitney, its new 16,000-square-meter manufacturing facility for its game-changing Geared Turbofan technology, the first in Asia, opened in early 2016. It will help to strengthen Singapore's capabilities in manufacturing advanced aircraft components.

With air traffic growing in the Asia-Pacific, airlines expanding and renewing their fleet, the prospects for Singapore's aerospace industry remain upbeat. Many look to Singapore for its strong engineering capabilities, a productive workforce, comprehensive intellectual property regime and pro-business environment.

As a leading aerospace hub in Asia, Singapore remains committed to further develop its infrastructure, capabilities and manpower competencies to support the future growth in MRO, manufacturing and R&D activities. The availability of expertise and ability to create the capacity to accommodate the growing demand for aerospace services will enable the industry to capitalise on opportunities from emerging markets and further boost Singapore's position as an aviation hub of choice.

As the aviation industry grows with complexity and sophistication, human capital development in aviation's specialised and technical industry becomes increasingly crucial. In building its own competencies and expertise in various areas of aviation, Singapore has learnt from its more advanced counterparts and has adapted solutions to suit its own special circumstances.

5.3. Aerospace engineering education & training

Singapore's ability to develop a world-class aviation and aerospace hub has always depended on its talent pool. As the aviation industry grows with complexity and sophistication, human capital development in aviation's specialised and technical industry becomes increasingly crucial. In building its own competencies and expertise in various areas of aviation, Singapore has learnt from its more advanced counterparts and has adapted solutions to suit its own special circumstances.

From the expatriate engineers who maintained early piston-driven aircraft at Seletar to the aeronautical engineers graduating from local universities, Singapore's skilled professionals have been essential to our achievements in aviation. Going forward, maintaining Singapore's success as a thriving global aviation hub will require the support of passionate individuals with even higher skill sets and education.

Ask any industry executive what key challenges they face and the answer will almost always be the same — manpower. This is true across all aspects of aviation, including aircraft engineering, technical support, services, aircraft management, airline pilots and air cargo logistics.

To meet the increasing demand for skilled manpower, the number of industry-specific educational programmes offered by local institutions has risen sharply in recent years. As a global aviation hub, Singapore recognises the importance of investing in human capital development, enhancing the skills of today's aviation professionals and nurturing the next generation to take the industry into its next stage of growth.

Today, in response to the rapidly growing needs of the aviation industry, Singapore's institutes of higher learning, universities and private training organisations offer a comprehensive range of academic, training and development programmes. There are courses for passionate youths pursuing aviation careers, aviation executives seeking to upgrade themselves and for mid-career professionals looking to join the dynamic industry.

Qualifications are also available at various levels — diplomas, specialised and postgraduate degrees and executive programmes — to feed the industry's demand for a host of qualified professionals to fill jobs.

The Institute of Technical Education (ITE), for example, offers courses in Aerospace Technology, Aerospace Avionics and Aerospace Machining. It is committed to nurturing and equipping students with technological, methodological and social competencies through hands-on training.

To further enhance student training, ITE established a cooperation framework with ST Aerospace under an MOU to work together on various aspects of training. This included setting up an aircraft airframe workshop at ITE, offering scholarships to students, providing on-the-job attachments for ITE staff and knowledge-sharing and development of joint certification courses.

Like ITE, Singapore's polytechnics offer a selection of aviation-related courses to reach out to a wide pool of aviation enthusiasts. They cover a variety of focuses and provide training for different jobs, such as apprentice licensed aircraft engineers, technicians and trainee engineers. The courses consist of classes laboratory sessions, site visits and work placements, which combine to instill a strong foundation in the fundamentals of aviation-related knowledge and equip the students with relevant practical skills as they prepare to enter the industry. They are also able to interact with industry practitioners and become exposed to emerging trends and technical knowledge.

For instance, Temasek Polytechnic offers three full-time diploma courses in aviation management and services, aerospace engineering and aerospace electronics, which all have a flight training option for students to obtain a Private Pilot Licence. The Temasek Polytechnic-Lufthansa Technical Training Centre also offers technical training and is opened to full-time students and working adults.

At the tertiary level, local institutions offer programmes that aim to mould ardent aviation students into professionals. In 2005, the School of Mechanical and Aerospace Engineering at Nanyang Technological University (NTU) was the first to offer the Aerospace Engineering course to university students in Singapore, with more than 500 graduates to date. More recently, it collaborated with DSO National Laboratories to produce Singapore's first locally built satellite, X-Sat, and its launch is one of the highlights of 2011. Students now have the opportunity to work with an actual satellite in space and NTU hopes that this will inspire the next generation of students and scientists to focus on engineering research and development.

The National University of Singapore (NUS) offers an aerospace specialisation within its School of Mechanical Engineering, which sees students taking subjects relevant to aeronautical engineers. In 2010, Airbus' parent company, the European Aeronautic Defence and Space Company (EADS), partnered NUS for a research and development project focusing on the fundamental study of flow control technologies for reducing drag around streamlined bodies, like aircraft. As part of the agreement, the NUS-EADS Internship Programme was launched to sponsor selected undergraduates from across disciplines, including Computing, Science and Engineering, to intern in one of EADS' research centres in Europe.

SIM University (UniSIM), meanwhile, has carved a niche for itself by offering flexible and customised education for working professionals who are either seeking a career switch or are already in the aviation industry and would like to broaden their skills. The chance for an education upgrade and increased competency in their work has drawn many mid-career and seasoned professionals through UniSIM's doors.

Professional training institutions like the Singapore Aviation Academy (SAA) — the training arm of the Civil Aviation Authority of Singapore — provide courses and training programmes for aviation professionals locally and internationally. Some of these are carried out under various training fellowship programmes as part of its mission to build human capital for the global civil aviation community.

The SAA also jointly offers graduate and postgraduate degree programmes with world-famous academic institutions like Cranfield University, Massey University, University of California, Berkeley and Embry-Riddle Aeronautical University. Through its four schools — School of Aviation Management, Aviation Safety & Security, Air Traffic Services and Airport Emergency Services — the SAA provides a variety of specialised aviation training programmes catering to the needs of industry and civil aviation professionals worldwide.

Certified programmes aside, the SAA organises forums, conferences, seminars and workshops for knowledge-sharing and networking within the international civil aviation community. These enable SAA to stay at the forefront of aviation knowledge and industry developments and forge alliances with leading aviation organisations and institutions worldwide.

Through the SAA and its partnerships, Singapore has created numerous platforms for the confluence of global leaders, top civil aviation officials, academics and leading industry practitioners to exchange views, debate and formulate new ideas for the advancement of international civil aviation.

As an International Civil Aviation Organisation (ICAO) Regional Training Centre of Excellence, SAA will continue to lead in the development and delivery of competency-based ICAO training packages.

Aviation companies are also seeing the value in aviation skills training and investing in state-of-the-art training facilities.

Home-grown SIA Engineering Company (SIAEC), an industry veteran for over 30 years, believes in the constant upgrading of employees' skills to keep up with the industry trends and demands. The company believes that gearing up its employee capabilities for the up-and-coming fleets like the A380 and B787 Dreamliner will ensure that SIAEC remains ahead in the value chain. The company's Training Academy, located at Loyang, boasts a technical library and state-of-the-art e-classrooms to support practical skills training for its employees and those of its subsidiaries, joint-venture companies and strategic customers and partners.

Another example is aerospace giant, Boeing. Over the past decade, Boeing has further expanded its presence in Singapore including the opening of a Boeing Flight Services training campus in Singapore, which is also its Asia-Pacific training headquarters. Boeing's training campus installed two new full-flight simulators — 777 and Next-Generation 737 — for training to support growing pilot training needs and increasing airplane deliveries in the region.

As aviation advances into the next century of commercial aviation, Singapore's dedication to develop aviation knowledge and thought leadership remains strong. It will continue to seek novel ways to work with key aviation's stakeholders to develop global civil aviation and help formulate and shape international policies and standards. By engaging minds and building a culture of continuous learning, Singapore will further build on its efforts to share knowledge as well as experiences and develop aviation human capital that will aid the industry in forging new paths and driving change.

5.4. Air defence

Singapore's defence components — the Republic of Singapore Air Force (RSAF), the Defence Science and Technology Agency (DSTA) and DSO National Laboratories (DSO) — play a critical role in the development and growth of Singapore's commercial aviation hub.

Like its civil aviation counterparts, the RSAF and its predecessor, the Singapore Air Defence Command (SADC) have grown in both scope and capabilities alongside Singapore's impressive economic progress. Widely recognised internationally as a small but potent, technologically advanced and professional air force capable of punching far above its size, the RSAF's transformation has been intertwined with that of commercial aviation from its earliest days.

From the formative days of colonial Singapore when Seletar Airport was established as Royal Air Force (RAF) Seletar in 1928 commercial aviation shared the airfield from 1930 until the opening of Singapore's first civilian airport at Kallang some seven years later.

When the Paya Lebar Airport was eventually opened, commercial aviation and the air force shared common aviation infrastructure once again.

But with independence and the announcement of the imminent withdrawal of all British troops by the end of 1971, plans for developing Singapore's defence forces were pushed into high gear. Formed in 1968, the SADC's immediate task was to set up the Flying Training School to train pilots. A number of local RAF technicians were released to join the fledging SADC where their experience working on fixed-wing RAF aircraft was invaluable to the nascent air force. Gradually, the SADC had its own pilots, flying instructors, air traffic controllers and ground crew.

Britain's former air bases — Tengah, Seletar, Sembawang and Changi — were handed over to the SADC, as were its air defence radar station and surface-to-air missiles. In 1975, the SADC was renamed the RSAF. A key development in those early years was the establishment — at the direction of Minister for Defence at that time, Dr Goh Keng Swee — of a number of defence-focused companies. Among these was the Singapore Aerospace Maintenance Company in 1975, whose first task was to refurbish second-hand US Navy Skyhawk aircraft for the RSAF. This company was later to become Singapore Technologies Aerospace, which, together with SIAEC, contributed towards Singapore's now substantial commercial aerospace industry.

Over the years, key defence-related institutions were established to meet the needs of the growing Singapore Armed Forces (SAF). A key institution, DSTA was created as a statutory board in 2000, under the Ministry of Defence (MINDEF) for greater autonomy and responsiveness to better address defence and security challenges. Its mission is to harness and exploit science and technology and provide innovative solutions to meet the defence and security needs of Singapore. By employing leading-edge technologies, DSTA continuously delivers new capabilities to enhance the operational capabilities of the SAF. Through DSTA, Singapore has earned the international reputation as a reference customer for defence acquisitions.

Its evaluation of major platforms and weaponry systems is closely watched and widely praised as being thorough and comprehensive. Over the years, DSTA has acquired advanced fighters, unmanned aircraft systems, helicopters, transport, and mission's aircraft to enhance the RSAF's operational capabilities. DSTA's was particularly recognised for their multi-disciplinary approach and ability to integrate diverse technology in different domains and deliver innovative and cost-effective solution.

DSTA also helps to build a strong community of engineers and scientists from the universities, research institutes, government and industry in the aerospace industry. DSTA has built up strong capabilities in systems engineering and systems integration which allows them to build innovative and first of its kind solution to meet the challenging needs of defence and security. Some of these engineering capabilities were later adapted for civilian use. One example was the quick development of the Infrared Fever Screening System (adapted from the military thermal scanners) to combat the spread of the Severe Acute Respiratory Syndrome (SARS) in 2003. Deployed at major checkpoints, including Changi International Airport, the thermal scanners was adapted from military thermal imager. The prototype was developed within days by DSTA engineers in collaboration with ST Electronics. The system made it practicable to screen large groups of people entering into Singapore, minimising the impact of SARS outbreak on the commercial aviation industry.

Chapter 7

Infocomm Technology

1. Introduction

In this day and age, it is hard to imagine a day without making a phone call to loved ones or business associates, or a day without e-mail, instant messaging, and the Internet. The importance of Infocomm Technology (ICT) to Singapore and the world is easy to see. A natural evolution of the telecommunications sector, Singaporean engineers have played an important role in its development in this country.

This chapter focuses on their efforts to bring about a more efficient system in Singapore, tracing its history and some technological solutions that were created in response to various needs and problems.

1.1. *History*

Before independence

It was during the colonial period that Singapore's telecommunication began to develop, so as to facilitate the conduct of trade and business. In 1879, telephones were introduced in Singapore with a 50-line exchange set up in 1881. The telephone network then was operated by the Oriental Telephone and Electric Company (OTEC).

In 1955, the British colonial government took over OTEC's operations due to the company's growing inability to keep up with subscriber's demand. It then established the Singapore Telephone Board (STB) to handle domestic telephone services while another entity, the Telephone Department, gained responsibility over trunk and international telephone services.

After independence: 1960s to 1980s

STB merged with the successor of the Telephone Department, the Telecommunication Authority of Singapore (TAS), in 1974. The new statutory board retained the TAS name and was formed in order to provide better and more efficient services. TAS merged with the Postal Service Department in 1982 in order to optimise resources. By this time, more than 500,000 telephone lines had already been installed, offering services such as the mobile telephony, International Direct Dialing (IDD), and paging.

1990s

The demand for telecommunications services grew rapidly during this period. In 1990, there was just slightly over one million fixed telephone subscribers in Singapore. This grew to 1.75 million by the end of 1998.

It was during this time that Singapore became the first country in the world to have a completely digital telephone network. Along with that, a nationwide broadband Integrated Services Digital Network (ISDN) was also put in place.

2000s

The telecommunications landscape evolved further in the 2000s, with Internet services coming into play, paving the way for the rise of the ICT sector. By 2009, there were three cellular phone operators in Singapore, serving more than six million cellular phones, and four major Internet service providers (ISPs) serving more than 4.8 million Singapore broadband users.

From merely having 50 telephone lines in the 19th century to the highly-connected nation that we are today, Singapore's telecommunication infrastructure has certainly developed over the decades. An idea of its growth in the past one-and-a-half decades alone can be seen in the data presented in Fig. 1 below. This would not have been possible without the support of the government and engineers from many disciplines.

2. Engineers' Contributions in Telecommunications and ICT

The world-class telecommunications and ICT infrastructure, together with extremely high Internet penetration rates that Singapore has today did not come

Year	2000	2005	2010	2015
Total Fixed line subscriptions	1,935,900	1,847,800	1,983,900	2,017,300
Fixed line population penetration rate	59.3%	43.3%	39.8%	36.4%
Total Mobile phone subscriptions	2,442,100	4,256,800	7,288,600	8,211,400
Mobile population penetration rate	74.8%	99.8%	143.6%	148.4%
Total fixed broadband subscriptions	70,933	622,708	1,337,891	1,461,074
Internet user penetration rate	36.0%	61.0%	71.0%	82.1%

Fig. 1. Fixed line, mobile, and broadband Internet subscribers and penetration rates, 2000–2015.
Source: IMDA and World Bank.

by chance. In the past few decades, Singapore's engineers have worked on new developments and innovations to improve the efficiency of our network and make it more accessible to the people.

2.1. *Improving efficiency of our network*

As the country grows and gets even busier, ICT will become a necessity to convey messages across the entire island and overseas. Various corporations work closely with engineers to bring out a better and more efficient system for easier and better communication. One particular area which has been under tremendous development is the Internet.

2.1.1. *Internet in Singapore*

In Singapore, there are about 9 million broadband Internet subscribers. There are three major Internet operators in Singapore, namely SingTel, StarHub, and M1. Together with smaller, newer providers such as MyRepublic, ViewQwest and Colt, they provide consumers with a variety of choices and plans to suit their needs.

Over the years, government has been promoting the usage of broadband Internet access. Internet access is readily available in Singapore, and it has brought convenience to all. As such, the infrastructure here has improved to allow better efficiency and performance, with efforts undertaken by various Singapore telecommunication companies as well as engineers.

Before the Internet, Singapore was the first country to launch an interactive information service to the public which included photographic images. This was known as Teleview, which was jointly developed by TAS and British telecommunications firm GEC-Marconi. This service started in 1987 and expanded in 1989.

Teleview was initially set up as a public service at the same time Singapore Telecom was formed in 1992 from the business arm of TAS. Subscribers connected to the Teleview service via a dial-up connection, initially by 1200–2400 bit modems (V22 Biz) and then later via 9600–14,400 kbit/s modems. A later development from Teleview provided an interfaced connection to the Internet — subscribers were given access to the Internet via a text-only terminal, e-mail was accessed by the text-based email client Pine and web pages were viewed by the Lynx text-based web browser. Teleview evolved to a full-fledged dial-up service for Internet access offered by SingNet, a subsidiary of SingTel.

The Singapore ONE project was formally announced in June 1996 in a government-led initiative to connect the island with a high-speed broadband network through various mediums such as fibre, DSL, and cable. By December 1998, Singapore ONE was available nationwide with the completion of the national fibre optics network.

In 1997, commercial trials for SingTel ATM-based "SingTel Magix" service were undertaken in March, before being launched in June. Also, in June, Singapore Cable

Vision commenced trials for its cable modem-based services, before being commercially deployed in December 1999. SingTel's ADSL service was subsequently rolled out on a nationwide scale in August 2000.

Current development

In January 2001, the Broadband Media Association was formed to promote the broadband industry. By April the same year, there were six broadband Internet providers. In October 2001, Pacific Internet introduced wireless broadband services.

In December 2006, Infocomm Development Authority of Singapore (IDA) introduced "Wireless @ SG." It was a part of the government's Next Generation National Infocomm Infrastructure initiative, and offered everyone free wireless access in high human-traffic areas, such as the Central Business District, downtown shopping belts like Orchard Road, and residential town centres. Initial access speed was 512 kbps, and has progressively been increased to 2 Mbps, and then to 5 Mbps in April 2016.

Also, in 2007, IDA launched the ambitious Next Generation Nationwide Broadband Network (Next Gen NBN) programme to provide the nation with ultra-high-speed broadband services. Deployment started in 2009, and by early September 2010, Internet service providers in Singapore rolled out Next Gen NBN service plans. The world's first, Next Gen NBN offers a pervasive, open-access network that provided competitively priced broadband speeds of up to 1 Gbps at prices comparable to ADSL and cable connections. As of 2013, Singapore Next Gen NBN achieved nationwide coverage with over 30 retail service providers and 10 underlying operating companies respectively offering retail and wholesale ultra-high speed broadband services.

On 1 July 2016, Prime Minister Lee Hsien Loong presented the Next Gen NBN programme team with The Institution of Engineers, Singapore (IES)'s Top 50 Engineering Feats award, cementing its importance in improving the lives of Singaporeans and the efforts made to arrive at this point.

2.2. *Improving standards of living*

Singapore is a well-connected society where citizens and businesses use communication technologies extensively to enrich their lives. To sustain this momentum, various organisations drive programs in key areas of ICT adoption and fine-tune the regulatory environment where necessary to support the people.

2.2.1. *Wireless @ SG*

Wireless @ SG is a wireless broadband programme. It aims to extend Internet access beyond homes, schools, and offices to public places. With that, people can enjoy free, both indoor and outdoor, seamless wireless broadband access with speeds of up to 5 Mbps at public areas. It allows users to access media-rich and interactive websites as well as use bandwidth-intensive applications such as video-streaming.

Whether it is for business, for school, or for entertainment, people can now have connection "on the move" while away from their home broadband. This not only provides people with a convenient way of connecting to the online information, but it also enables people to get connected with one another.

In April 2013, the next phase of the Wireless @ SG programme was launched to improve user experience and drive the adoption of innovative consumer and enterprise services. Enhancements to the network include:

- Progressively higher access speed.
- Easier login to the network through SIM-based authentication for mobile devices.
- Innovative services, such as targeted advertising, office communication suite, and Software-as-a-Service (SaaS).

Although Singapore is not the first country with wireless network in the public, it has one of the best communication systems in the world.

2.2.2. *Next Gen NBN — toward an infocomm-enabled future*

The Next Gen NBN is another example of how Singapore continues to further improve her telecommunications network for the next generation. It provides a nationwide ultra-high speed broadband access of 1 Gbps and more to all physical addresses including homes, schools, government buildings, businesses, hospitals, and NBAPs (Non-Building Access Points). This infrastructure is a critical national enabler that spurs the development of new knowledge-based sectors, such as research and development, interactive digital media, and creative industries. It catalyses development and deployment of innovative interactive digital services, such as Cloud Computing and high-definition videoconferencing, to homes, schools, and businesses.

The Next Gen NBN reinforces the status of Singapore as an infocomm hub and opens new doors to economic opportunities, business growth, and social vibrancy.

2.2.3. *Heterogeneous network (HetNet)*

Today, Wi-Fi and cellular networks operate in silos. Networks have different range and data rate, and they do not readily share information such as traffic volume and latency with one another. Even though networks for mobile devices can be switched either manually or automatically, the networks lack the intelligence to switch to best suit the needs of the user. To improve the standard of living, the network should be accessible anywhere, anytime, using any devices. As such, different networks need to be integrated and operate as a unifying heterogeneous network ("HetNet").

In 2016, when the Infocomm Media Development Authority (IMDA) was formed, it undertook exploration of the concept of a heterogeneous network (HetNet) as a strategy to enable everyone and everything to be always connected via the best available network, any time, any place, and at high speeds.

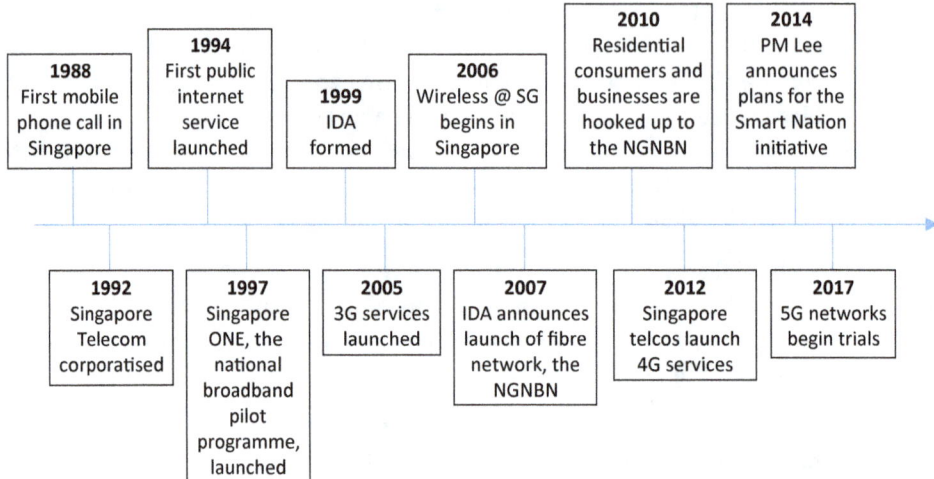

Fig. 2. Some key ICT dates in Singapores history. *Source*: IMDA, NUS and Smart Nation Programme Office.

The specific aims of IMDA's HetNet concept are:

- Mitigate capacity crunch through better utilisation of spectrum and infrastructure resources. More importantly, the HetNet program should help mobile network operators (MNOs) to lower their total cost of ownership which eventually translates to lower cost of service for users.
- Optimise overall network capacity to deliver consistent quality of experience, as the HetNet programme will likely improve the quality of service (QoS) of networks by distribution of load across multiple wireless access networks. As a result, users will always be connected to the network which can meet their requirements.
- Facilitate intra-operator roaming to build a resilient infrastructure as the HetNet program will bring about seamless roaming within the networks of each MNO, thereby improving network resiliency. The primary focus would be to facilitate roaming capabilities across different radio-access technologies (RATs).
- Identify new business opportunities and models as the HetNet program will provide an environment for MNOs and solution providers to uncover new revenue streams through innovative network-sharing techniques (e.g. wholesale model) or services (e.g. premium QoS offerings). It will also provide opportunities for joint participation from different industry sectors and different MNOs to interoperate and build stronger business relationships.

2.3. *Telecommunications go green*

2.3.1. *Kim Chuan Telecommunications Centre 2 ("KCTC-2")*

Green building and construction

KCTC-2 is the Next Generation Green Data Centre. Designed for high performance and energy efficiency, it is SingTel EXPAN's latest state-of-the-art data centre, delivering highly secure and reliable IT infrastructure services.

As one of the biggest data centre in Singapore with 150,000 sq ft of co-location space, KCTC-2 provides the latest data centre technologies with a full suite of managed service offerings to customers looking to outsource their IT infrastructure and operations.

Besides providing high performance and reliable services, SingTel, together with its contractors and engineers, contributed to environmental sustainability through the inclusion of eco-friendly and energy-efficient features in KCTC-2, which was awarded the BCA Green Mark Gold rating.

2.3.2. *Green services*

As the nation moves toward a greener future, greener technological solutions have been developed. In SingTel EXPAN, green services include:

- Thermal analysis of the data centre space to identify ways to optimise energy and cooling efficiency through computational fluid dynamics (CFD).
- Power reports and tools for real-time visibility into energy consumption levels.
- Virtualisation of servers to reduce physical servers' footprint and utilise space effectively.
- Use energy-efficient servers that provide better efficiency and allow for more servers in the same space compared to conventional servers.

2.4. *More solutions, products, and services*

As telecommunication becomes more and more important in the current society, telecommunicating solutions, products, and services must improve and upgrade continuously to meet the demand of users.

ST Engineering is a Singapore-homegrown engineering group specialising in innovative solutions and services in aerospace, electronics, land systems, and marine sectors. Within the electronic sector, many solutions related to telecommunications have been developed.

2.4.1. *Simplex satellite telemetry devices*

Installed on mobile assets, the tracker receives location signals from a GPS satellite and relays its current location through low-earth orbit satellite networks at regular intervals to the end-user. The asset's movement can thus be tracked and be monitored if it is within its predesignated zones. The sensor can be wired up through its general-purpose input/output (GPIO) utility for event-triggered application. It can use power line or be fully battery-powered.

Designed to the needs of many users, the device is able to perform tasks including:

- Geo-Fencing
- Remote Environmental Monitoring
- Asset Tracking

- Panic Bottom/Emergency Status Reporting
- Remote Fault Reporting
- SCADA/Messaging

2.4.2. *Mobile Earth Station (MES)*

The Mobile Earth Station (MES) integrates the power of a high-quality satellite communications system with the mobility of a durable trailer to provide reliable field communications. Designed for easy transportability and mobility, the system meets commercial land, air, and sea transport restrictions and specifications. Efficiently mobile, the MES is also designed for safe and rapid deployment.

Fig. 3. Mobile Earth Station (MES). *Source*: ST Electronics.

2.4.3. *Modems*

ST Electronics' latest satellite modem ASM 5800 series incorporates all of the features of the ASM 4800, and adds 8PSK and 16QAM modulation modes, a standard Reed-Solomon and IBS multiplexer, and higher data rates. The multiplexer's overhead channels provide the capability for orderwire, AUPC, and remote modem control to every modem. The ASM 5800 series is available in IF versions for 70 MHz, 140 MHz, and L-Band applications and comes with various data rates, 5 Mbps, 10 Mbps, or 20 Mbps options.

2.4.4. Satellite broadband communications

As a leading manufacturer and solution-provider of satellite broadband communications, ST Electronics has been equipping both public and private organisations with groundbreaking applications for two decades. Manned by a team of dedicated engineers, equipped with advanced and comprehensive facilities sited in high-technology manufacturing premises, ST Electronics designs, develops, and markets reliable and versatile communication products and solutions to customers in more than 100 countries. This number is set to grow with the advent of seamless end-to-end broadband wireless solutions for various communication needs, business applications, and platforms.

2.4.5. Sprint Software Programmable Radio

The SPRINT Software Programmable Radio (SPRINT SPR) is designed to deliver voice and data (with maximised battery life) over the most demanding and mobile *ad hoc* environment. It is specially developed with programmable hardware platforms to provide network-centric capability, flexibility, upgradeability, and versatility suited for dismounted military/paramilitary use in urban terrain, unmanned platform systems, unmanned sensor systems, etc.

3. Our Future and the Smart Nation Initiative

With the advent of digitalisation, it is not just our ICT and telecommunications sectors that are affected. In fact, new possibilities that enhance our way of life, our productivity, and even how we play have also opened up.

Taking the lead in adopting and experimenting with new technologies, in the name of creating a better living space and opportunities for all in future Singapore, the government has embarked on the Smart Nation initiative, launched by Prime Minister Lee Hsien Loong in 2014. It is currently headed by Dr Vivian Balakrishnan, who is also the Foreign Affairs Minister.

Many programmes and initiatives, such as the testing of autonomous vehicles by LTA, the implementation of smart home solutions in Yuhua estate, and the opening of a centre for healthcare-related assistive robotics at Changi General Hospital, have since been launched under the auspices of this initiative.

IoT, or the Internet of Things, is also an important development arising from digitalisation. Besides smart home applications (e.g. remote monitoring, security and comfort management), agencies like LTA, NEA, HDB, JTC, PUB have plans to deploy various smart sensors to monitoring the environment, traffic flow, utility consumption, security, and so on, so as to enable policymakers to be better-equipped to make decisions that will improve the lives of Singaporeans.

The government has also encouraged the use of robotics and manufacturing methods such as 3D-printing, together with further development of cloud computing, artificial intelligence systems, and the like, so as to be able to 'tech-up' and transition to Industry 4.0 without being left behind.

A major concern with all this increased interconnectivity would be cybersecurity. The criminal element, or anti-establishment entities, will no doubt seek to destabilise the order by exploiting technological loopholes. As such, the field of cybersecurity will also feature heavily in our ongoing march towards our Smart Nation goals.

What will the future bring? Driverless vehicles, digital-enabled homes, classrooms, workplaces, near-instantaneous, long-distance communication, and maybe more. Considering all these, it is then evident that we will need engineers — not just in the traditional sense, but cross-trained in several disciplines, with a dash of digital knowledge and data analysis thrown in.

Chapter 8

Offshore & Marine

1. Introduction

The offshore and marine industry has played an important role in the transformation of Singapore from a fishing village in the last century into one of the world's major offshore and marine, and shipping hubs during the last 50 years. Today, Singapore is a global leader in high specifications jack-up rig construction, floating production storage and offloading (FPSO) vessel conversion, and Liquefied Natural Gas (LNG) carrier repairs. It is also a major builder of semi-submersible platforms and premier ship repair centre in the world.

1.1. *The early days*

The development of the industry to its present status can be traced to more than 150 years ago when the first dry dock was built. It provided Singapore with a head start into ship repair, as it was the only repair dock between China and India.

Shipbuilding activity also has its roots in humble beginnings catering mainly to river and port traffic. It started with the construction of simple wooden barges and tug boats by a number of shipyards whose activities did not range beyond the region.

The lead was extended following Singapore's independence in 1965, when the industry was identified for development and expansion to create employment for the nation. The industry gained traction in the late 1960s with a steady inflow of investment by both the public and private sectors. The first commercial shipyard was set up in 1963, followed by the privatisation of the government's dockyard and conversion of the naval dockyard into a commercial ship repair yard in 1968. Swamps along Geylang River were cleared to make way for the first marine industrial estate — the Kallang Marine Industrial Estate.

The next few years saw the rapid increase in ship repairing activities and shipbuilding orders, resulting in more investments in infrastructure and facilities and expansion of capacities to meet growing demands. New shipyard facilities were also set up in the newly created Jurong Industrial Estate and Tanjong Rhu Basin. By 1975, Singapore had become the largest repair centre for ocean-going vessels.

2. Staying Ahead of Competition

The discovery of oil in Southeast Asia in the 1960s provided a second wind for the industry and set the rig industry booming. By 1970, five rig builders had established in Singapore, turning it into a centre for rig building, second in importance only to Houston.

The dramatic increase in shipping activities, ship repair and shipbuilding orders created a huge demand for marine-related support services. The growth of a comprehensive marine support industry resulted in a critical mass of ancillary services. By the 1980s, Singapore had developed into a comprehensive hub for marine and offshore products and services. Ship repair jobs also increased in complexity to include jumboisation, modification, conversion and life extension of ships.

Despite the slump and being branded as a "sunset industry" during the economic recession in the 1980s, the industry managed to turn itself around and transformed into a sector of growth. The offshore and marine industry has remained viable and become even more relevant to the Singapore economy. Today, it is an integral part of the Singapore maritime cluster, contributing to the nation's role as a leading international maritime centre.

By the 1990s, the industry had moved on to higher value and technically more complex and sophisticated jobs. Companies adopted new technologies to advance their engineering and production processes, and developed proprietary designs and creative solutions to offer their clients. Industry players also rationalised to counter competition from new, low-cost shipyards in China and the Middle East. Following a series of mergers and reorganisation, two major marine groups — Keppel Offshore & Marine (Keppel O&M) and Sembcorp Marine — emerged with subsidiaries specialising in ship repairing, shipbuilding, rig-building and offshore engineering.

In ship repair, Singapore is a world leader for LNG carrier repair and life extension work. LNG carriers are high value assets and the maintenance of LNG vessels is a field reserved for a select few with proven track records, given the ship owners' stringent standards on quality and safety. Singapore is also the regional centre for cruise ship repair, upgrade, refit and revitalisation.

3. Conversion Specialist

Conversion and modification of ships was a major source of revenue for the shipyards as it draws on their wealth of engineering expertise and technical experience. Since the 1980s, when general cargo ships were converted to cattle carriers for the Middle East, Singapore shipyards had undertaken projects to modify cable ships to cruise ships, lengthen container ships and dredgers, using their proven engineering skills, procurement and construction expertise and strong organisational capabilities. The types of conversion projects undertaken here in Singapore ranged from car ferry to dynamic positioning accommodation and repair vessel, scientific research vessel into integrated ocean drilling vessel, fishing trawler into seismic research vessel, container ships to livestock carrier, and tanker to FPSO.

Fig. 1. An example of an FPSO. The Capixaba was a fast track conversion completed by Keppel Shipyard in 2006.

Before the turn of the century, Singapore has become the conversion capital for FPSOs with 70% share of the global FPSO conversion market. Its shipyards are capable of providing a full range of conversion requirements from engineering to construction, outfitting, integration and commissioning. In a fast-track conversion, where concurrent engineering is undertaken, Singapore shipyards can complete the project in less than 12 months (Fig. 1).

With its accumulated engineering experience and project management skills honed from FPSO conversions the industry is able to tap on to new growth markets presented by the global trend towards green shipping. Singapore has since progressed to convert a LNG vessel into the world's first LNG Floating Storage and Re-Gasification Unit (FSRU). A FSRU is a cost-effective and flexible solution to receive and process LNG shipments.

Apart from FSRU conversions, Singapore has also moved into another specialised segment of the conversion market, the Floating Liquefied Natural Gas Vessel (FLNGV). FLNGV conversion offers cost-effective alternatives to land-based LNG facilities and new-build FLNGVs, allowing gas to move faster to market.

4. Going Big on Offshore

"Yards in Singapore have reached a very high level of competence... I expect the rigs (built here) to outperform anything in the industry." Those were the words of

Mr Claus Hemmingsen, board member of AP Moller Maersk, on the construction of offshore rigs by Singapore shipyards.

The words summed up the expertise and position of Singapore in the global rig construction stage. Singapore's rig building took off in 1969 when Far East Levingston Shipbuilding delivered the first jack-up rig, J W McLean. By 1980, Singapore's five rig builders then, had among them secured a total of 31 contracts, propelling it into the premier spot as the top jack-up rig builder in the world.

Over the years, Singapore shipyards have built up their capability with proprietary and engineering solutions to sustain its lead. Today, Keppel O&M and Sembcorp Marine continue to keep the Singapore flag flying with their range of jack-up (Fig. 2) and semi-submersible rig (Fig. 3) projects.

5. Customised Shipbuilding

Apart from drilling rigs, the shipyards are also building accommodation and other floating platforms and fabricating topsides and modules for offshore facilities and oilfields. In recent years, they have also ventured into drillship construction for deep water operations.

Once dismissed as a sector with a limited future, Singapore's shipbuilding has come into its own. The local shipbuilders have carved out a niche as specialist builders for smaller and customised vessels. In the past, shipbuilding involved primarily tugs, ferries, supply boats, bunker tankers and container feeder ships.

In the 1980s, the shipbuilders moved up the value chain to build container ships, cement carriers and product tankers. The range has since expanded to include anchor-handling tugs, cable-laying ships, pipe-laying vessels, RoRo (roll-on/roll-off) vessels, ice-breaker ships, dredgers, offshore supply vessels, fire-fighting, rescue vessels and naval ships.

6. Embracing Technology

Technology development is pivotal to Singapore's continuing success as a premier international marine and offshore centre. To stay ahead of competition, companies invested heavily in technology and in research and development (R&D) so that they can continue to innovate.

The industry's search for better processes began in the 1980s to curtail the need to employ ever-increasing numbers of workers to take on the increasing volume of work. The advent of computers in the 1980s saw the industry embracing CAD/CAM technology. Shipyards adopted CAD/CAM, numerically controlled burning machines and semi-auto welding machines to expedite work and enhance efficiency.

Their efforts were complemented by industry-initiated programmes. Urged on by the government, five major shipyards came together to standardise work procedures for ship repairs. 11 standard procedures were drawn up covering work in areas such as turbines, centrifugal pumps, cargo blocks, condensers, boilers and main engines. Mechanisation was introduced in work process to improve productivity.

Fig. 2. B Class jack-up rigs, constructed by Keppel FELS for a Mexican customer.

Both Keppel O&M and Sembcorp Marine have invested considerable sums on R&D to maintain their market leading positions. Sembcorp Marine Technology was set up in 2007 to sharpen the company's competitive edge in offshore and marine technology. The R&D centre focuses on new product development and process innovation.

In the same year, Keppel O&M established its Keppel Offshore and Marine Technology Centre (KOMTech) to focus on developmental technology. The new centre strengthens the R&D initiatives undertaken by Keppel O&M's technology

Fig. 3.   A semi-submersible rig being constructed by Sembcorp Marine.

units — the Offshore Technology Development, Deepwater Technology Group and Marine Technology Group.

Government bodies such as Agency for Science, Technology and Research (A*STAR), Economic Development Board (EDB) and Maritime and Port Authority of Singapore (MPA) have lent a helping hand and partnered the industry to reshape the offshore and marine landscape. Together with research agencies like the NUS Centre for Offshore Research and Engineering (CORE) and Ngee Ann Polytechnic's Centre of Innovation (COI) for Marine and Offshore Technology, they worked with shipyards and marine and offshore companies to transform Singapore into a global marine and offshore hub.

7. The Keppel Experience — The Industrialisation of Singapore and the Birth of Keppel

It was an entrepreneurial decision — Hive off the dockyard department of the Port of Singapore Authority, which had always been just a servicing adjunct, and turn it into a full-fledged business of its own. It contained repair ships from all over the world, not just those that visit the port, and had a turnover of S$35 million a year. The aim? Make it S$50 million a year in five years' time.

And so was born Keppel Shipyard (Pte) Ltd (Fig. 4), wholly owned by the Singapore government, on 23 August 1968. The S$50 million target was set by its

Fig. 4. Keppel Shipyard today.

first chairman, and in effect, its founding father, Mr Hon Sui Sen, who was then also the chairman of the Economic Development Board (EDB).

Mr Hon and his colleagues in government dared to dream big dreams: they envisioned Singapore as the Asia's largest ship repair centre after Japan. At hand were three major shipyards — Keppel Shipyard, Jurong Shipyard Ltd, a joint venture between the EDB and Ishikawajima-Harima Heavy Industries of Japan, which was set up five years earlier; and Sembawang Shipyard (Pte) Ltd, wholly owned by the Singapore government, which would by the end of the year take over and run the dockyard in the naval base. Britain had in January decided that it would withdraw from its bases in Singapore within three years.

The year 1968 saw not just the birth of Keppel, but of Sembawang, Neptune Orient Lines, (the national shipping line), and the Jurong Town Corporation (JTC), which became the centre for an industrialised, export economy.

The late '60s were good years for the ship repair business. The shipyards doubled their output from S$64 million to S$120 million in 1968, and turned in S$140 million in 1969. In 1970, it grossed a record turnover of S$200 million. Oil exploration in the region and the expansion of oil refining activity in Singapore were responsible directly or indirectly for many of the orders.

In 1970, the then-Prime Minister Lee Kuan Yew challenged the shipyards to make Singapore the biggest ship repair centre between the Persian Gulf and Japan. Singapore was the fourth largest port in the world, but it had yet to become the fourth largest shipbuilding and ship repair centre in the world.

"There is no reason why we should not be if we take full advantage of the present fortuitous circumstances and build soundly on it," he said. *"If we can keep up the speed and quality of repairs, establishing a firm solid reputation, then even after these fortuitous circumstances, we would have got well established and will continue to have a flourishing business."*

Mr Lee Kuan Yew

Keppel shared in the boom. Its turnover went from S$33 million in 1969 to S$44 million in 1970 to S$60 million in 1971.

7.1. *Building Singapore into a ship repair centre*

Up till the late 60s, Singapore did not figure at all in the world shipping community as a ship repair centre. What it had was a busy port, a dockyard as an adjunct to the port, a naval dockyard and some small yards. The new Singapore government saw a potential market of S$500 million a year by the end of the '70s. It set up Keppel and Sembawang, after having set up Jurong Shipyard with Japanese partners.

By 1974, its turnover was already a whopping S$720 million. Overnight, Singapore became not only a ship repair centre that mattered, but it also became effectively the most important ship repair centre between the Persian Gulf and Japan.

The many more private shipyards numbered over 50, and they included two major yards, the S$90 million Hitachi Zosen Robin Dockyard and the S$80 million Mitsubishi Singapore Heavy Industries. These made up the Big Five, together with Keppel, Sembawang, and Jurong. The shipyards were clustered in three areas — in the north around Sembawang Shipyard; in the south in Tanjong Rhu, and Keppel; and in the west at Jurong. In Tanjong Rhu, there were a number of smaller yards: Weng Chan, Kwong Soon Engineering, Eagle, Singapore Slipway, and Kall Teck. In 1965, there were 25 yards with some 5,000 employees. By as early as 1971, the industry's workforce was 18,000 strong, more than one-eighth of the total number in the manufacturing industry.

As early as 1971, Keppel took its first step into diversifying its business by buying up 40% of Far East Levingston Shipyard (FELS), which dealt mainly with offshore work.

The mid to late '70s presented a turbulent patch. The oil crisis caused by the Middle East war toward the end of 1973 caused a worldwide recession. Nearly 30 million dwt of tankers were "mothballed," and new orders were cancelled in 1975.

The larger yards also took this opportunity to consolidate by tightening production, cutting down on waste, and utilising the workers more efficiently. This was also the time for them to diversify by going for smaller vessels and conversion jobs.

When the upturn came toward the end of the decade, the shipyards were leaner and fitter. The first two years of the '80s saw spectacular performances: the turnover was S$1.9 billion in 1980 and S$2.4 billion in 1981.

In 1982, the Port of Singapore Authority (PSA) invested S$216 million to convert the conventional berth at Keppel Wharves and the Victoria and Albert Docks to two container berths. This was part of its S$500 million plan to expand the Tanjong Pagar container terminal. Keppel had to give up the two old docks, in exchange for PSA's Pulau Hantu. Keppel would rename the island Pulau Keppel and make it a repair facility, with a floating dock.

7.2. *The recession*

The storm came in 1982, and lasted for a good five years, shaking up the ship repair industry for good. The yards began cutting down their prices, by as much as 25%. In 1982, the number of vessels coming in for servicing or repairs began to drop.

The five major yards — Keppel, Sembawang, Jurong, Hitachi Zosen Robin and Mitsubishi — started competing for smaller tankers and cargo vessels of 5,000 dwt, and effectively snatched away the businesses from the smaller yards.

For the first time in its short history, ship repair suffered a decline in earnings in 1982. Total turnover fell by 21%, from S$1,088 million to S$860 million.

The inevitable shake-up came in 1984 and 1985. Hundreds of workers were laid off. Medium-sized yards such as Sing Koon Seng and Weng Chan were folded up, and American giant Marathon LeTourneau shipped out. Mitsubishi pulled out of ship repair, closing down its 400,000 dwt dock. It had lost S$15 million in 1983 and

S$10 million in 1984. The 12-year-old joint venture between the Singapore Government and the Mitsubishi group of Japan would concentrate on plant engineering instead.

During the rig boom in the early 1980s, 300 rigs were built worldwide. By the end of the 1990s, 74 of the 82 rig building yards around the world ceased operations. Singapore saw three out of five rig building yards cease operations as well. They were Marathon LeTourneau, Bethlehem Singapore, and Robin Shipyard. The fourth, Promet Shipbuilders, drastically scaled down its operations and maintained only its key personnel for rig repair and conversion jobs. It did finally resume rig building activities, but only after a management buyout in 1997. Its name was changed to PPL Shipyard (PPL). In 2001, SembMarine, which wanted to venture into the rig building business, acquired 50 per cent of PPL. In 2003, SembMarine increased its stake to 85%.

7.3. The start of Keppel FELS and rig specialisation

The company that not only remained active but went on to expand its operations during the prolonged slump was Far East Levingston Shipbuilding (FELS). It is one of the three main companies that today make up Keppel O&M.

To survive, FELS took its first major calculated gamble in 1985, with US$40 million at stake. It entered into a joint venture with US rig designer Friede & Goldman (F&G) and a French partner to develop and build a new rig in anticipation of a buyer. FELS was the only company in the world that was building a rig in 1986.

In 1990, only three contracts to build rigs were awarded worldwide and FELS managed to clinch all three. During that period, FELS decided that building jack-up rigs would become its core business. It therefore had to enhance its technological capabilities.

7.4. Developing its own jacking system

Another bold but necessary move FELS made was the creation of its first design and development company, Offshore and Technology Development (OTD) in 1994.

At that time, there were only two companies in the world supplying jacking systems — one American and one French. FELS was totally reliant on them, not only on quality control but also pricing and delivery schedules. The management at FELS saw an urgent need to secure good supplies of critical rig components such as jacking systems.

It approached the Singapore National Science and Technology Board (which has since evolved into the present-day A*STAR) for a grant under the Research and Development Assistance Scheme. With the grant, it set up OTD and its development of a jacking system was also successful. The breakthrough came when GlobalSantaFe decided to use the FELS jacking system on one of its rigs, Galaxy II.

A major milestone was reached in 1997. The offshore industry was in a downturn and rig designer Friede & Goldman (F&G) wanted to raise cash. FELS, which was renamed Keppel FELS that year, took the opportunity to buy the design and drawing rights for two F&G jack-up rigs, the Mod V and Mod VI.

The purchase made OTD a rig designer. It now had the full rights to modify and improve upon these designs. To enhance its rig design expertise, Keppel FELS formed an alliance with Bennett & Associates (BASS), a rig design firm.

Having acquired rig designs, Keppel FELS was able to make in-roads into the US drilling market. The Keppel FELS jack-up rig would eventually become known as KFELS A Class — a harsh-environment jack-up rig fully designed by the Keppel team at OTD, Keppel FELS, and BASS.

7.5. *The oil price boom*

After a deep slumber that lasted 20 years, the offshore and marine industry began to stir once again in 2004. Throughout that year, the world economy underwent strong recovery. Demand for oil and gas grew by more than 3%, the fastest since 1976. As a result, crude oil prices rose to record levels, sparking renewed interest in offshore oil exploration and production.

The pick-up in market demand benefited all companies in Keppel O&M, not just rig builder Keppel FELS. Keppel Singmarine started 2004 with eight orders in February and by July, three more Anchor Handling Tug Supply (AHTS) contracts were secured. With these contracts, Keppel Singmarine's order book grew to 23 vessels.

In addition, there was strong demand for Floating Production Storage Offloading (FPSO) vessel conversions (Fig. 5), and Offshore Support Vessels (OSVs).

7.6. *Execution excellence*

1987 would be the first time that FELS would bid directly for a contract to build an oil rig by an oil company — the Conoco platform project. Oil companies were particularly stringent in their requirements but FELS completed the rig on time and under budget and enhanced Keppel's reputation for executing complex jobs on time, safely with high quality and at competitive prices.

In 2008, Keppel FELS became the first Singaporean company to receive the Manufacturing Excellence Award (MAXA), which is accorded to Singapore-based companies that have achieved world-class manufacturing standards.

In 2013, Keppel FELS delivered 21 rigs, setting a Guinness world record for the most number of rig deliveries by a company in a single year.

7.7. *Leader in jack-up rig designs*

In 2000, Keppel FELS launched the KFELS B Class jack-up drilling rig as a cost-efficient rig for benign waters. It was designed to exceed the standards of the best ultra-premium rigs of the day, with improved productivity, enhanced safety, and added independence from supply lines and environmental conditions.Chiles Offshore

Fig. 5. FPSOs undergoing conversion.

signed contracts for the first two KFELS B Class rigs, one to be built in Singapore and another in USA.

The new design quickly gained market acceptance. This led to the development of the KFELS Super B Class rigs, capable of drilling to 35,000 feet below the seabed.

In 2012, the qualities of the KFELS Super B Class and KFELS B Class Bigfoot were combined to produce the KFELS Super B Class Bigfoot design for Transocean.

7.8. Keppel's investment in deepwater technology

In 2006, there was a clear shift toward the construction or conversion of semi-submersibles. Thus, after its integration in 2002, one of the first investments made by Keppel O&M was to establish a separate Deepwater Technology Group (DTG) as a research and development company for semi-submersible rigs and other floating structures.

It is the only shipyard group in the world with its own deepwater semi-design capabilities and its own suite of proprietary deepwater solutions.

Solutions developed by DTG for deepwater drilling include the Extended Semi-submersible or ESEMI. This has a second set of pontoons — which can be retracted or lowered as necessary — to provide greater stability in the face of ocean waves.

DTG also designed the ESEMI II, an adaptation of its ESEMI design. The ESEMI II is designed to reduce riser dynamics and improve the fatigue life of the Steel Catenary Risers.

One form of technology that has made deepwater rigs a practical proposition is the dynamic positioning system or DPS. This uses satellite technology to maintain the position of the rig, which is fitted with eight thrusters that automatically move it back into place when it drifts. The rig is continuously keeping itself in place. Such technology is necessary because at such depths, it is technically impossible to moor a rig with anchors.

Keppel FELS was among the first companies to adopt the system and the first to have it commercially tested by a marine testing group. The technology is now routinely used not just in semi-submersible rigs built by Keppel FELS, but also in many specialised offshore support vessels built by Keppel Singmarine.

Keppel O&M's successful DSS^{TM} Series of semi designs were developed with its design partner GustoMSC. The first of this series was the DSS^{TM}20 rig, Maersk Explorer, which was delivered to Maersk in 2003.

In 2013, Keppel FELS delivered 21 rigs and set a new record for the most number of rig deliveries by a company in a year.

7.9. Beyond rigs

As the offshore and marine boom continued into 2006, there were more than 100 rig building projects in the bidding, design and planning stages in that year — compared with 50 new orders in 2005 and just 9 in 2004. Equally strong was the demand for FPSO conversions, OSVs, and ship repair services.

Since Keppel Shipyard's first offshore conversion project in 1981, when it delivered one of the world's first converted FPSO units, the company has built up a strong track record of more than 100 FPSO, Floating Storage and Offloading (FSO), and Floating Storage and Regasification (FSRU) conversion and upgrading projects as at the end of 2015.

Keppel Singmarine started to build its first ice-breakers and FSO vessel, while Keppel Shipyard began the conversion of the world's first FSRU and has started a new business in completion work for new drillships. Keppel FELS broke new ground in its traditional stronghold of jack-up rigs as well, when a new class of extra-large, high-performance jack-up rigs was introduced, putting the value of jack-up rigs on par with that of semisubmersibles.

Keppel O&M's specialised shipbuilding division continues to design tugboats in Singapore through Marine Technology Development (MTD) and constructs them mostly in China and the Philippines. MTD, set up in 2003, focuses on research into offshore support vessels and specialised units.

The Keppel O&M Technology Centre (KOMtech), established in early 2007, underscores Keppel O&M's commitment to enhance its investments in long-term

Fig. 6. The two ice-breakers, *Varandey* and *Toboy*, built by Keppel Singmarine. The two ships currently operate in the Barents Sea, clearing the path for vessels to travel through the frigid waters there.

research. Some of these investments will extend beyond the group's boundaries. It started with an initial funding of S$150 million for five years.

KOMtech also collaborated with the Energy Research Institute @ NTU (ERI@N), which focuses on sustainable energy and energy efficiency/infrastructure, and the A*STAR Computational Resource Centre (A*CRC), which provides high-performance computational resources and training courses in computational fluid dynamics simulations.

7.10. *The ice edge*

In 2008, Keppel Singmarine delivered *Varandey* and *Toboy* (Fig. 6) to LUKOIL, Asia's first two ice-breakers for the Arctic. Built in compliance with the Russian Maritime Register of Shipping's standards, the two ice-breakers are capable of cutting through solid ice over 1.5 m thick and operating in extreme temperatures as low as −45°C.

Since then, Keppel Singmarine delivered several other firsts, including the Caspian region's first ice-class FSO Yuri Korchagin in 2009, which incorporated new in-house designs by MTD.

Keppel O&M also has the capabilities to build mobile ice-resistant offshore drilling units and ice-capable jack-ups. In 2012, KOMtech and ConocoPhillips collaborated to jointly design the first-of-its-kind ice-worthy jack-up rig.

7.11. Research into LNG technology

In 2015, Gas Technology Development (GTD) was set up to champion and develop the Group's solutions across the gas value chain from the gas fields to the end users. For instance, Keppel Shipyard is currently undertaking the world's first-of-its-kind conversion of a MOSS LNG carrier into a Floating Liquefied Natural Gas Vessel (FLNGV), which is expected to complete in the second half of 2017.

Keppel's FSRUs enable customers to bring their gas to market without relying on land-based terminals. As specialised shipbuilders, they also designed a range of small-scale LNG bunkering vessels and barges that can deliver LNG along shallow water rivers, coastal areas, and inter-islands.

Furthermore, Keppel O&M's 65-ton bollard pull LNG dual-fuel Azimuth Stern Drive (ASD) tug jointly developed by KOMtech and MTD, won the Outstanding Maritime R&D and Technology Award at the 2015 Singapore International Maritime Awards.

In addition to the LNG-fuelled tugs which represent solutions in engine conversion, Keppel's Floating Storage Regasification and Power (FSRP) barges and vessels provide downstream solutions in power conversion and can be installed quickly and economically.

Chapter 9

Health & Safety

1. Introduction

Health and Safety Engineering (HSE) is the application of engineering concepts, regardless of engineering disciplines, to meet health and safety goals. HSE is a multi-disciplinary area, and it is an important foundation for Workplace Safety and Health (WSH). This chapter captures the growth of HSE and WSH in Singapore in the past 50 years. WSH is a fundamental part of any engineering discipline, and engineers always balance safety and health with production goals and resource constraints. Thus, engineers have to be creative to achieve the different goals while ensuring that the system that they design and build are safe.

Engineers are an important stakeholder in the development of WSH in Singapore, and they have contributed as policy-makers, legislators, designers, technical experts, investigators, inspectors, authorised examiners, qualified persons, professional engineers, chartered engineers, trainers, workplace safety and health officers, industrial hygienists, and so on.

However, WSH, as well as workers' well-being, tend to be neglected during a nation's infancy. Hence, the quest toward nurturing WSH in Singapore was extremely arduous, and engineers had to learn from painful accidents to prevent recurrence and improve the safety performance of engineered systems. In the 1920s, rules and legislations modelled after British labour laws for occupational safety and workmen compensation were passed in an attempt to raise the importance of health and safety in the workplace. In subsequent decades, the WSH landscape made significant improvements by overcoming difficulties that arose in every new era. To become a nation of WSH excellence, new engineering initiatives and strategies were introduced proactively.

The pioneer generation established the foundation for WSH in the industry despite Singapore being a young nation. As time progressed, government bodies, engineers, industry, public interest society and professional alliances came together and enabled Singapore to achieve successive WSH goals.

Different professions contributed to the growth of WSH in Singapore. One of these professions is engineering, which included civil and structural, mechanical, electrical, and chemical engineering. In fact, engineers had always been at the core of Singapore's WSH regulatory framework.

The push for WSH proved to be advantageous for the economy since safe work environments led to an improvement in productivity along with a reduction in lost man hours due to injuries, ultimately boosting investor confidence and leading to greater economic and job prospects for our nation. It also raised social benefits as individuals viewed safety and health with utmost importance.

Reaping the benefits from the call for action by Prime Minister Mr Lee Hsien Loong in the National WSH Campaign 2013, the industry has since saw a decrease in death rates to 1.8 per million employed persons in 2014 along with increased numbers of safety champions to adopt the ongoing efforts of developing a safety culture in their workplace.

This chapter retraces the WSH and HSE history from Singapore's independence to the present.

2. HSE in its Infancy (1965–1974)

When Singapore gained independence in 1960s, industrialisation was the main driver for economic survival. To increase employment rates, the government introduced a large-scale industrialisation plan to turn Singapore into an industrial hub. Following the success of the Jurong industrial area, industrialisation development boomed, creating abundant jobs, particularly in the marine and oil and gas industry.

While the economy thrived, novel issues, especially with regard to WSH, came up. Some of the initial challenges faced in WSH were due to inexperience since workers were either oblivious to the fact that they were working in unsafe conditions or were untrained in operations of advanced equipment. Consequently, death and accident rates hiked by three and seven times, respectively, between 1963 and 1970.

Furthermore, dangerous work conditions coupled with harmful work settings resulted in excessive cases of silicosis in labourers working in confined spaces. The respiratory occupational disease was the "number one industrial killer" in Singapore then. To tackle this issue, the government sought assistance from the industrial health experts of the International Labour Organization (ILO) to carry out a study on occupational health issues in Singapore in March 1966.

Besides stressing the importance of supervision and law enforcement during the industrial periods, the government also recognised the need to coach and connect with the workforce. Therefore, small-scale factories were urged to display safety posters while larger ones were obligated to form safety committees, which provided a platform for engineers to contribute to WSH. In addition, an occupational safety seminar was also organised by the National Safety First Council for the first time since its establishment on 1 July 1966.

Later in 1969, the formation of the Occupational Safety and Health Committee under the National Trades Union Congress created motivation toward healthier and safer working circumstances. The committee was full of enthusiasm in imparting a higher sense of awareness among the public and general body of workers.

In order to better manage the workload and simplify WSH-related efforts, the Industrial Health Unit (IHU), led by Dr Chew Pin Kee, was established in 1967 by the Ministry of Labour (MOL) and Ministry of Health (MOH). The IHU was responsible for studying reportable occupational ailments, provision of guidance on factory-related health and environmental problems, and participation in studies and training. In a survey on granite quarries in Singapore carried out by Dr Cressall, an occupational health specialist from ILO, it was observed that the silicate dust counts onsite greatly surpassed the maximum permissible exposure limit specified in international occupational health standards. To alleviate the problem, quarry occupiers were ordered to implement dust control measures, including engineering systems.

In the same year, the Sand and Granite Quarries Regulations were enacted to effectively curb issues of silicosis in the industry. In addition, the Abrasive Blasting Regulation passed in 1974 also proved to be effective in controlling silicosis by banning sand usage as abrasives in blasting. A series of regulations such as the Building Operation and Works of Engineering Construction Regulation 1971 were also enacted to protect safety and well-being of workers. A number of these regulations required design and certification by engineers.

Although regulations and prosecutions provided impetus for change, the government recognised the fact that many areas of Singapore's industrial safety were still lacking. Hence, a study group on Accident Prevention in Shipyards was formed in 1973 to increase awareness and provide solution to the industry. This study group saw active participation from engineers in both the public and private sectors. Its research concluded in March 1975 with a four-volume report which was presented to MOL. In the report, the shipyard industry was observed to be the greatest contributor in accident occurrence and it was recommended that the industry adopt greater preventive measures, safety legislation, a code of practice and a standardised system for accident reporting. In May 1975, an advisory committee was formed to aid in the execution of recommendations made. Engineers played a significant role in the development and implementation of these recommendations.

The WSH journey toward greater ownership started in the 1970s with the launch of the first National Industrial Safety and Health Campaign in 1972 followed by a campaign on silicosis using mobile exhibits in 1973. Besides these, MOL also disseminated monthly newsletters called The New Worker to allow workers to stay relevant on occupational safety and health news. Furthermore, the Safety Officer Training Course (SOTC) was also launched in 1973 to mentor individuals to become safety officers. Many of the trainers in the SOTC were engineers, who contributed their engineering expertise in hazard identification, control, monitoring and management.

3. Growth of HSE (1975–1984)

During the period from 1975 to 1984, the size of the industrial workforce increased from 257,300 workers in 1975 to approximately 408,700 in 1982. Amidst busy industrial activity, occupational mishaps and high fatality rates continued to be the leading sources of worry. As highlighted by the study group on Accident Prevention in Shipyards, the shipbuilding and repair industry contributed to 30% of total accident occurrences in 1975, many of which could have been avoided if the industry had adopted attentive and organised measures toward inculcating safety awareness. Met with these situations, the government was certain that regulations alone were insufficient to stimulate change, and that implementation coupled with engagement, plus regulation and outreach had to be adopted.

Thus, the government increased the tempo of engagement efforts through national campaigns that introduced unfamiliar aspects of occupational safety to the public and raised awareness among workers of high-injury risk industries.

The first national campaign was introduced by then-Minister of Labour Ong Pang Boon on 24 September 1975. The three-day campaign reached out to more than 26,000 shipyard workers, supervisors, management, and stakeholders. The highlight of the campaign featured a mobile exhibition based on trips to 21 Singapore-based shipyards showcasing pictorial displays of the do's and don'ts on board vessels, safe and unsafe work procedures, and remedial actions to take in the event of mishaps. In addition, interactive tools such as televised forum on Accident Prevention in Shipyards, interviews between MOL officials and shipyards, poster and slogan competitions, and safety movies were provided. Furthermore, the ministry also prepared a three-day safety orientation course to aid shipyard workers in getting acquainted with the permit to work system and safety precautions.

Apart from coaching the workforce via educational campaigns, the government also realised the need for greater coordination with the industry. With this aim, a 15-man Advisory Committee chaired by Prof Ang How Ghee was formed in 1975. The committee was responsible in helping the shipbuilding and repairing industry to embrace safety culture and providing information and assistance on the execution of suggestions made in the report by the study group.

In March 1976, a Management Workshop on Safety in Shipyards was formed to emphasise the risk associated and precautions required for confined spaces. Motivated by the knowledge gained, the shipyard representatives started providing basic safety resources and having safety policies and an in-house permit to work system.

Half a year later, the Safety Consultancy Group was founded to provide advisory assistance to shipyards on WSH and to aid in the implementation of suggestions recommended by the Advisory Committee. Many of its members were engineers. To confirm their commitments, shipyard contractors were urged to endorse the agreement to take preventive measures against workplace accidents. This call for action was supported by nearly 95% of the contractors by November 1977.

Apart from the shipyard industry, an advisory committee was later established in the construction industry after observations of a steep increase of 141 accidents between 1979 and 1980. The advisory committee worked together with the Factory Inspectorate in organising safety orientation programs for tertiary students, construction workers, and supervisors. Through collaborations with MOL, the committee later facilitated a Safety and Construction workshop, which was introduced in March 1980 by Mr Ong Pang Boon.

As a result of the endeavours of the committee, more industry members were able to gain insights into safety and health in work environments, converting many formerly passive participants to active WSH advocates.

While staying involved with industry leaders and workforce, MOL made an effort to ensure that WSH standards stood strong and updated through constant assessments of WSH legislation.

The impetus for self-regulations in WSH became obvious when the Factories (Safety Committee) Regulations was enacted in 1975. Under the regulation, safety committees made up of appointed management representatives and selected staff within factories that employed more than 50 people would have to carry out regular plant inspections, monthly WSH meetings, on-site accident investigations, and raise awareness for safe work practices. The requirement for management and staff attendance in meetings on safety issues would ensure better communication between the top brass and line workers. Moreover, closer interactions would also help in the prevention of accidents as erroneous behaviour can be promptly recognised and corrected.

Within the same year, the Factories (Qualifications and Training of Safety Officers) notification was also released. This notification fully explained the duties of a safety officer. Collectively, the regulations strengthened participation and supported WSH standards in the workforce.

Besides the enactment of new regulations, it was also acknowledged that there was a need to change mindsets. While existing WSH efforts focused on a preemptive approach by eradicating safety threats promptly before the occurrence of fatal accidents, the new approach focused on routine inspections. The result of this new mentality was the launch of the Industrial Hygiene Monitoring Programme in 1983.

Under this programme, high-risk worksites such as petrochemical complexes, asbestos factories, and granite quarries were required to carry out inspections every quarter by the Industrial Health Division (IHD). As a means of measuring the effectiveness of WSH control measures, a worksite must keep documentary proof of measures made.

Another concern made was with regard to safeguarding the health of workers in dangerous work settings. Thus, compulsory health assessments were arranged to help them diagnose and treat workplace-induced illnesses early.

MOL also carried out a close monitoring of the construction industry through special enforcement operations involving many engineer inspectors to remove careless contractors. 122 contractors were found liable for fines or prosecution right off the bat. Further in 1980, 48 particularly dangerous worksites were imposed with stop work orders. Since stop work orders would lead to additional time and money incurred, it proved to be an effective deterrence toward recalcitrant offenders.

By joining the forces of engagement and education together with proactive and harsh measures, the government was better able to fight safety negligence and remain vigilant.

In 1980s, Singapore started constructing MRT tunnels, marking an important milestone for the industry and HSE. To undertake this massive infrastructure project, engineers were sent overseas to places like Japan for training, so as to increase competency in safety engineering for tunnelling works and not compromise on it at all.

A working committee, comprising of engineers from the then-Factory Inspectorate and industry partners, was then established to provide lessons on occupational safety and health for all individuals handling tunnelling works with the aim of ensuring their health and safety. In a bid to ensure effective communication to all stakeholders of the MRT project, two separate versions of safety guides were disseminated to the MRT contractors and construction workers, respectively.

Then, in October 1984, the MRT Construction Hygiene Monitoring Programme was introduced for the evaluation of control design, execution of site inspection, preparation of guidelines, and supervision of worksites. This programme provided a platform for IHD to meet up with contractors before commencement of tunnelling works, so that precautions were taken to mitigate possible health hazards.

Apart from these, a safety competition suggested by Mr Arthur Scott-Norman along with two other judges Mr H H Ho and Mr K S Lee was conducted to heighten knowledge of WSH among contractors.

4. Evolution of HSE alongside the Industry (1985–1994)

As the economy started to thrive in the 1990s, the industries became more differentiated with more technologically advanced jobs. With the industrial sector starting to undertake more sophisticated and larger assignments and introduction of sunrise industries, the need for complex engineering and equipment intensified. This in turn led to an increase in WSH risk. Therefore, the industry and government needed to pursue necessary engineering and management actions to guarantee safe usage of heavier, and more advanced equipment.

During the late 1980s, the government began constructions of key infrastructure for the nation. The most essential enhancement to the transportation system started in 1983 with the construction of MRT tunnels, lasting for four years before the first stations were launched in 1987. Amidst the period of inclination toward change, a series of catastrophic fire outbreaks and explosions took place in the shipyard industry between 1992 and 1994 causing high fatalities and injuries. These painful lessons resulted in a drive for higher WSH engineering and management standards.

As the characteristics of industrial projects changed, pressure vessels and lifting equipment started to come into usage. Pressure vessels, for example, are dangerous as they can cause injuries or death to nearby workers if they fail. Therefore, strict supervision of the design, production, and usage of these vessels was crucial. This made the importance of HSE even more prominent.

To reduce the heightened hazards associated with the use of such tools, the Ministry initiated legislations to mandate the inspections by certified inspectors before usage.

To facilitate independent and easy employment of certified inspectors, the Department of Industrial Safety (DIS) made use of engineering technologies to improve industry safety. By 1990, the online, text-based Teleview system could be used to provide data on certified inspectors that factories could employ for inspecting their lifting equipment, pressure vessels, and other machines used. In 1992, the Department upgraded the online system to enable inspectors to carry out data entry for pressure vessels using their private workstations.

On 1 January 1994, the Factories (Crane Drivers and Operators) Regulations that was proposed in the previous year came into being. Under this regulation, every tower and mobile crane operator had to be taught to follow safety requirements. Not only that, the regulation also mandated the need for assigning lifting supervisors to oversee lifting operations. Additionally, safety instruction courses for lifting supervisors were also introduced.

The other project that was closely monitored by the Singapore government was the construction of MRT tunnels led by the Mass Rapid Transit Corporation (MRTC) in 1983. Since safety was made the top priority for all stakeholders and contributed to the key performance indicators of contractors, near-misses were recorded and lessons learnt were communicated to workers to prevent future recurrences.

In addition, a permanent safety manager was engaged by the MRTC who would be tasked to adhere to a robust structure and convey information promptly to the appropriate project directors for the project. Through stringent checks and balances, safety was given due care despite hastening the building process.

Since the project team comprises employees of different nationalities, mutual agreement was tough as individuals had different work norms. Hence, management depended on a reliable safety structure throughout the organisation as this arrangement aided in bracing synergies and coordination in the project and safety teams.

As time progressed, the Ministry began to recognise the need to safeguard the health and safety of workers of other industries. Consequently, the Department of Industrial Health made attempts to form a legislation to mandate workers whose job scope entails selected risk factors for medical assessments by authorised family doctors. This arrangement guaranteed that the workers were healthy enough to work and could avoid overt occupational diseases through early detection.

After the Factories (Medical Examinations) Regulations came into act in 1985, confirmed cases of occupational disease increased by more than 749.

On 15 March 1986, Hotel New World collapsed leaving 17 injured and 33 dead. Within less than a minute, the entire structure broke down into ruins. Although help was dispatched within 8 minutes from the occurrence, rescue operations were hindered as the rescue teams were inexperienced in handling such incidents then. Fortunately, the rescue operation was able to pick up speed after the tunnel engineers involved in the construction of the MRT construction were called in. Their aqua jet cutting tools, life detectors and infrared imagers greatly assisted in rescue efforts.

A week after the tragedy, a Commission of Inquiry was appointed to carry out investigations on the collapse. Among the findings reported in the final report of 16 February 1987 were problems of additional roof installations and superficial maintenance work on structural cracks which further worsened the underlying problem of poor structural design of the building.

As a result of the tragedy, new laws were passed which mandated a compulsory maintenance checks on all buildings with public access every five years, and implementation of double checks on structural plans and calculations by professional engineers. Furthermore, professionals are also required on-site to supervise structural works; random spot-checks were carried out; and owners were also required to test the piles and structural materials used.

In 1991, DIS continued to make improvements in reducing hazardous working circumstances by introducing the need for all factories to be registered before operations. The DIS also implemented a safety management program to encourage organisations to make safety their priority, to allocate resources to strategise safety promotion movements, and to assess its safety performance.

In 1994, amendments to the Factories (Building Operations and Works of Engineering Construction) Regulations were made where it defined the obligations of contractors to engage permanent and temporary safety supervisors and separate safety auditors to regulate worksite safety standards.

The Ministry of Labour then noticed that prescriptive-based WSH management was insufficient to totally prevent accidents. Therefore, it began to launch training courses for employers and staff to expand their knowledge on individual tasks and competencies in accident deterrence.

As for the oil and petrochemical industries, the Safety Orientation Courses (SOC) that were introduced in 1991 were especially intended for employees working in manholes and other confined spaces due to the dangerous nature of their work. The course also taught the attendees on how to prevent accidents and inspired them to inculcate a culture of keeping an eye on each other with regard to WSH.

To solve the issues brought about due to diverse home cultures affecting the safety attitudes of foreign labourers here, the SOC scheme was introduced in 1993 to the construction industry. Foreign labourers were obligated to finish the course before getting their work permits and to attend refresher courses every two years as a mandatory requirement for work permit renewal. Courses then were constantly updated and taught in a variety of languages so that all attendees could understand.

Furthermore, the Occupational Safety and Health (Training and Promotion) Centre (OSHTC) also provided many WSH lessons to companies' management staff and professionals. These were customised to allow participants to learn more about the practical methods of detecting and assessing safety and health risks and to develop a safety management system for managing diseases and accidents.

5. HSE in the New Millennium (1995–2004)

This decade marked the start of the digital era where information technology was greatly capitalised in all developing countries. As the nation progressed into the new millennium, the economy began to face unfamiliar difficulties and developments. As the economy matured, higher skilled and better paying jobs were created.

Despite a reduction in employees in the manufacturing industry in 2000, the injury and death rate stagnated, thus requiring the government to persist in solving the issues faced so that safety and health of workers in these sectors could be guaranteed.

Continuing from the past decade's attempts to promote WSH, MOM and the industry focused on the persistent problems of sectors experiencing high death and accident rates through the use of carrot-and-stick policies.

In 1998, a partnership between the Department of Industrial Health (DIH) and Ministry of Finance (MOF) was formed to establish tax incentive schemes for noise and chemical hazard control in factories to reduce their economical encumbrance. The way these schemes worked was that they allowed the factories 100% depreciation claims for the first year of spending on effecting control measures for noise and chemicals. Additionally, meetings were also organised to publicise solvent and noise hazard controls and related health and safety risks at work.

Another problem area that the Ministry had to tackle was dealing with construction firms with lacklustre safety standards. Hence, in April 2000, MOM launched the Debarment Scheme targeted at raising construction safety, and introducing the Demerit Points System (DPS) to discourage contractors from neglecting safety in construction. Consequently, contractors possessing poor safety records would be debarred from hiring overseas employees. As time progressed, the DPS went through multiple reviews so as to stay updated to the ever-changing WSH standards.

In 2001, the division started announcing a series of campaigns and enforcement efforts to decrease the incidence of chemical over-absorption in the ink, plastics manufacturing, and printing industries. As a result, there were 24 less such cases as compared to that in 2000.

March 2003 marked the start of the Severe Acute Respiratory Syndrome (SARS) outbreak, killing 33 people before being eradicated four months later in July.

MOM was quick in its response towards the outbreak as it cooperated with the Ministry of Health (MOH) to manage the problems brought about by SARS to the workforce, through prevention and containment measures, warranting trade stability and providing coaching and assistance to organisations and firms to handle the epidemic.

Brochures and notices were distributed in four languages (English, Malay, Tamil, and Mandarin) to foreign labourers to provide information and measures against SARS that all workers could understand regardless of nationality. In addition, OSHD was also involved in contact-tracing among various ministries. As a result of their relentless efforts, Singapore was declared SARS-free by the World Health Organisation on 31 May 2003.

Subsequently, the collapse of the Nicoll Highway on 20 April 2004 was a tragedy that left an impact in Singapore's construction history. The incident was a result of the failure in recognising the severity of the buckling and sagging of beams leading to the eventual collapse of earth onto the excavation site for the Circle Line (CCL) construction, killing four workers.

The incident highlighted the importance of safety following the loss of precious lives which can never be recovered and raised the sense of urgency. Within two days, a Committee of Inquiry was tasked to investigate the reasons for the collapse and recommend measures to prevent future such incidents from happening.

By 30 August 2004, 133 witnesses had responded to the inquiry while an interim report was created. The full report was then submitted to MOM on 3 September. Following the official response to the interim report announced on 13 September, the government acknowledged the extensive suggestions highlighted, which included requirements for formulating risk management, safe procedures, and compelling pledge toward safety.

To restore public confidence, actions taken to embrace safety were announced. BCA and LTA's Building Control Unit then performed complete evaluations on its 14 CCL sites.

Where possible, additional steps were taken to avoid overstressing of materials. A timely warning system was also established to enable workers to detect predictable hazards. Based on the committee's suggestions, contractors would have to get authorisation from a professional engineer or qualified person from LTA to ensure that only experienced personnel could be deployed for specific tasks.

The sufficiency of emergency procedures was another problem that was highlighted in the interim report. To consider this problem, MOM inspected LTA's deep excavation sites and found that although safe excavation processes were in place, there was a lack of distinct instructions resulting in unfamiliarity with potential incidents arising from excavations.

In response to this problem, MOM and LTA's safety department gathered their Registered Safety Officer (RSO) to formulate distinct standards stating situations requiring emergency evacuations with LTA. Further to the other initiatives raised, the committee also emphasised the need to look at the overall situation.

Following the wake-up call from the Nicoll Highway incident, the WSH reformation took place. In 2005, the new WSH framework was established in support of greater ownership of WSH outcomes. The changes included mandatory requirements to engage specialists to handle important areas of geotechnical work such as soil investigation and instrumentation.

For example, Accredited Checkers (AC) are needed to carry out evaluations and site inspection works as well as review of trigger levels and interpretation of data while Professional Engineers (PE) are accountable for monitoring and instrumentation of the Temporary Earth Retaining System (TERS) in deep excavations, as well as data interpretations, and specification of review trigger levels. Besides the requirements for specialised professionals to be engaged on-site, they would also be required to attend meetings on design and construction-related issues to ensure that the building works progress safely and with quality. The subsequent decade then saw an era of major improvements as the Ministry worked together with the industry and its partners to foster greater WSH excellence.

6. Reformation of HSE (2005–2014)

This decade began with the unveiling of the new WSH framework. With the concerted efforts from multiple entities, WSH landscape saw progression amidst fast changes.

Trade unions related to the National Trade Union Congress (NTUC) were supportive of stronger teamwork among WSH stakeholders. Inspired by the constant discussion among the government, unions, and employers, the tripartite alliance saw the Singapore National Employers Federation (SNEF), Ministry of Manpower (MOM), and NTUC come together to collectively recognise the strengths and inadequacies of the industry, as well as what it needed to enhance WSH across the board.

During this time, the tripartite partners collaborated in promoting WSH as a career, coaching, re-tooling job specialisations, supporting equal employment opportunities and advancing work customs. This translated to increased economic attractiveness for the field and general improvements in WSH.

The journey for the new framework reformation started in early 2005 following the ministerial educational tour to Europe to comprehend and achieve wider perspectives on the nationwide WSH frameworks implemented in various countries. The Institution of Engineers, Singapore (IES), together with other industry associations participated actively in the education tour and contributed to the reform.

Led by the former Minister for Manpower and Second Minister for Defence Dr Ng Eng Hen, the delegations began understanding numerous elements of those countries' WSH framework including:

- The position and/or organisational arrangement of councils, industrial bodies, safety professionals, unions, and government agencies which influences on guaranteeing WSH.

- Independent frameworks such as the rewards and impetuses for industry to exercise impendency.
- The procedure and arrangement for evaluation of safety standards to guarantee their relevance with technological and industrial progressions.

After returning from the trip, the delegation gathered their discoveries to create the new WSH framework.

On 10 March 2005, the new framework became the turning point for Singapore's WSH journey, switching from being regulatory to one that supported more ownership for WSH consequences. The modified framework emphasised on three beliefs, namely, lowering hazards through reduction or eradication of risk creation by stakeholders, mandating industry to assume ownership of WSH outcomes through performance-based approaches, and avoiding accidents using heavier punishments for inferior safety administrations.

Backing the new framework, WSH 2015, strategised by both MOM and the WSH Advisory Committee (WSHAC), was created to evolve the nation into a country known for top WSH performance.

As a result of the success of WSH, the fatality rates in the workplace was lowered to 2.2 per 1 million employed persons in 2010, surpassing the government's aim to achieve workplace death rates of 2.5 by 2015. This success could be attributed to the industries' commitment in bolstering WSH and closer relationships with the government.

As supporters of WSH, the industry leaders adopted and carried out WSH fully within their companies over the past decades.

Industry engagement on WSH issues was further developed with the establishment of the WSHAC in 2005, bringing together 14 industry leaders and 4 industry-specialised committees, serving dual functions of guiding MOM on WSH regulations and standards and garnering higher industry support and contributions to maintain WSH quality. IES was one of the key members of this committee.

The WSHAC was later transformed to a fully developed council in April 2008 and was named the Workplace Safety and Health Council. It cooperates with MOM to strategie plans to improve WSH quality. Through numerous schemes, the council advocated compliance with WSH regulations and, for individuals, embracing and welcoming a deeply-rooted WSH culture.

In 2006, the WSH Act, which served as the chief legal means backing the new WSH framework, was enacted. The Act urged the implementation of required competency to detect and alleviate threats before its occurrence. Hence, safeguarding the safety and well-being of every worker became a common duty. The Act also propagated the notion that WSH responsibilities should not be limited to employers or WSH officers, and defined clearly the responsibilities of every stakeholder.

By 2008, the six phases of the WSH Act were introduced. In 2011, the Act was expanded to include all workplaces.

Another basis of the new WSH framework was the idea of risk management (RM), which is a crucial means for the detection and evaluation of possible hazards

Fig. 1. Then-Deputy Prime Minister and Minister for Finance Tharman Shanmugaratnam at the launch of the National Workplace Safety and Health Campaign in May 2014, where RM 2.0 was introduced.

as well as execution of control measures. Under the RM process, engineering control is always preferred as compared to administrative measures and personal protective equipment.

In 2006, the WSH (Risk Management) Regulations was implemented to require all employers to carry out risk assessment and execute measures to eradicate or lower workplace hazards. MOM and WSHC updated this with the introduction of RM 2.0 in 2014 (Fig. 1) to achieve three key purposes, so that organisations would view RM as a vital area of concern and as a part of their core values:

- Directing efforts on practical exercise to ensure identification of hazards and implementation of control measures.
- Lowering hazards at its origin and upward control measures must be executed as primary actions. Otherwise, other actions must be taken to safeguard the health and safety of employees.
- Organisations should adopt an overall approach on how personal risk factors can be integrated with traditional WSH risks in a holistic manner to ensure the safety and health of their employees.

Apart from helping organisations on WSH adoption, nurturing a WSH-competent workforce formed part of the WSH agenda. With the contributions from

MOM and the Workforce Development Agency (WDA), the joint WSH Professionals Workforce Skills Qualifications (WSHP WSQ) was introduced in April 2008 to provide WSH professionals with the opportunity to raise their competency in managing workplace hazards.

Aside from the construction industry, the marine sector was also provided with 13 trade-specific courses under the marine industry trade-specific WSQ framework. The industry also mandated new foreign workers to take on Shipyard Safety Instruction Course.

Also, in the metalworking industry, all foreign workers had to pass the Worker Safety and Wellbeing Test and show proof of attendance for the Metalworking Safety Orientation Course in order to remain hired. This has been in effect since 1 October 2009.

Moving forward, to realise Singapore's goal to become a WSH leader globally, the WSH Institute was launched in 2011 as part of the WSH 2018 strategy.

Although the goals for WSH 2015 were accomplished in 2009 as a result of the commitment from every member of the WSH landscape, the plans for new resolutions were already in the pipeline. In April 2008, the new national WSH strategy was declared by PM Lee targeting to reduce workplace fatalities to less than 1.8 per million employed persons by 2018. The new aim would place Singapore alongside world leaders of safety such as Sweden and the United Kingdom.

WSH 2018 aims to achieve four core strategic results:

- Reducing WSH incident rates.
- Making safety and health a key concern in firms.
- Ranking Singapore as a centre for WSH excellence in the sharing of knowledge globally.
- Creating an all-encompassing and progressive WSH culture locally.

7. HSE for the Future (2015–Future)

The shift from heavy industrialisation to knowledge-based enterprises resulted in the introduction of new and emerging trades. The diverse workforce, along with an ageing population, requires the management of advanced engineering techniques and technologies to maintain our competitiveness.

Despite rapid changes, engineers continue to be committed to the mission of providing engineering systems and solutions to achieve functional goals while ensuring workplace safety and health.

The transition from a prescriptive to performance-based approach did not come easy. Although various people and organisations believed strongly in the success of WSH, few of them were sceptical about its purpose.

Despite additional endurance and time-based requirements, commitment toward WSH eventually led to heartening outcomes.

In the process of WSH reform, the industry and government saw substantial improvements, with fatality rates successfully lowered to 1.8 per 1 million employed

Fig. 2. Mr Tan Chuan-Jin (third from left), then-Minister for Manpower and Dr Amy Khor (second from right), then-Senior Minister of State for Health and Manpower, at the launch of the WSH Campaign 2015.

persons in 2014, surpassing its original target in 2018. With these accomplishments, the industry and government next worked toward achieving Vision Zero.

On 24 April 2015, then-Minister for Manpower Tan Chuan-Jin disclosed a new objective for the WSH programme: Vision Zero, an attempt at instilling a mentality that occupational disease and injuries can be avoided through a pro-active approach (Fig. 2).

It was first introduced in the WSH stakeholders' dialogue in April 2013 to receive feedback from industry leaders on the viability of the movement. The primary result reflected that more than 90% of them were agreeable to adopting Vision Zero. Further, in 2014, additional suggestions for accomplishing this were made during the fourth meeting of the International Advisory Panel (IAP). In addition, the IAP highlighted a crucial stand, which was to view Vision Zero as a journey instead of a destination. Along with the WSHC and WSH Institute, the OSHD started rolling out initiatives to realise the Vision Zero mindset.

In the building industry, with its persistent injuries and deaths, Vision Zero became a wake-up call. During the Construction WSH Leadership Summit in 2015 (Fig. 4), more than 300 enterprise leaders vowed to lower death rates to less than 1.8 per million employed persons by 2018. This meant that the industry had to lower its death rates by a quarter annually for the next three subsequent years.

To reach the aim outlined in Vision Zero, the construction industry WSH Action Plan 2015 was developed to define strategies for greater WSH accomplishments in the built environment. It looked at how things could be done better in the various components of a project, such as training, cultural induction, leadership, procurement and Design for Safety (DfS). Also, the WSH Action Plan included enterprise leaders' commitment toward personal ownership of WSH and that toward devising solutions to avoid workplace injuries and ill-health.

Gazetted in August 2015, the WSH (Design for Safety) Regulations was enacted in 2016, outlining the roles and responsibilities of all parties to a project. Projects with a value of more than S$10 million are subjected to this regulation. To support the execution of DfS, MOM aims to train 1,000 DfS experts by 2018 and provide revision to existing DfS courses.

With these wide-ranging efforts put forth, the construction industry will be at its best position to accomplish its aim in the long run, making even more substantial improvements in WSH.

For the oil refining and petrochemical industry, substantial investments consist of buildings categorised as major hazards installations (MHIs) as they deal with huge amounts of toxic and flammable substances. The chances of major MHIs incidents are usually very slim, thanks to the presence of strict WSH management systems. Nevertheless, accident involving MHIs of any scale can be disastrous. Therefore, the maintenance of high MHI reliability levels is critical to workers, the public, and the economy.

In May 2013, MOM piloted a multi-agency study mission to the United Kingdom, Germany, and the Netherlands to enable the team to learn from global trailblazers in executing WSH within the MHIs. Following this trip, an inter-agency team was formed to assess the present MHI regulatory framework to achieve better management of health, safety, and environment hazards within the sector (Fig. 3).

Consisting of NEA, SCDF, EDB and JTC, the team sought to make the following revisions:

- A set of MHI Regulations to be enacted by the first half of 2016.
- National MHI Regulatory Office to be launched as the one-stop regulatory administrator for NEA, SCDF and MOM.
- The industry would be granted a year to prepare and improvise before the new regulations become effective in 2017.

This has resulted in the Safety Case Regime framework, a government policy initiative. In support of it, the Professional Engineers Board, working in conjunction with MOM, NEA and SCDF, introduced the Professional Engineer (Chemicals) qualification. PEs with this qualification will play an important role in the industry with the implementation of the new MHI regulations.

Besides accomplishing WSH developments locally, Singapore is also an enthusiastic member of the regional WSH landscape.

Fig. 3. A major hazards installations (MHI) study mission was conducted by an inter-agency team; subsequently a taskforce was set up to review Singapore's existing MHI framework.

Fig. 4. Minister for Manpower Lim Swee Say (third from left) and WSH officials at the Construction WSH Leadership Summit 2015.

From 3 to 6 September, Singapore hosted the XXI World Congress on Safety and Health at Work 2017. Organised by the International Labour Organization, International Social Security Association, as well as the Ministry of Manpower, it is the largest occupational safety and health in the world and the first-ever to be held in Southeast Asia. The theme of the congress, *A Global Vision for Prevention*, is suitably aligned to support the Vision Zero mindset.

Over the four-day event, a range of activities and programmes targeted at nurturing WSH best practices and norms were held. These included technical sessions, symposiums, poster presentations and industrial visits. An *International Media Festival for Prevention*, showcasing films and multimedia productions from all over the world about WSH was also held to raise greater awareness about this critically important issue.

8. Conclusions

The achievements today depended largely on the perseverance of our forerunners and generations of WSH advocates who had aimed for improved workplace safety and health (WSH) standards. Due to their selflessness, Singapore is now a "WSH-conscious nation" with robust tripartism and distinct ambitions.

However, this is only the start as it requires more than just ambitious foresights to keep the developments going.

Adopting the right mentality would be helpful toward achieving success. Apart from monitoring statistics, all employees within an organisation should embrace and exercise a WSH mindset daily. This includes adopting a culture of keeping an eye out for each other's well-being, health, and safety. When the culture is deeply rooted in all workplaces, promises can be realised and a true understanding of safety would win the hearts and minds of all — a result that everyone should seek.

The journey toward exemplary WSH needs continuous vigilance, contribution, and involvement from everyone. Our engineers are more than committed to this journey.

References

Building and Construction Authority. (2005). *Safety Actions after Nicoll Highway Incident.* Retrieved February 17, 2017, from https://www.bca.gov.sg/aboutus/others/ar05_11.pdf

National Library Board. (2011). *Hotel New World Collapse.* Retrieved March 4, 2017, from http://eresources.nlb.gov.sg/infopedia/articles/SIP_783__2009-01-02.html

The Straits Times. (1987). *How to Avoid Another Hotel New World Disaster.* Retrieved from http://eresources.nlb.gov.sg/newspapers/Digitised/Article/straitstimes19870329.2.3

Conclusion: Engineers for the Future

Background

Technology consistently drives humanity forward, from simple improvements in our standards of living to overcoming immense global challenges. The engineering of transnational pathways, sustainable energy, international transportation, and global communication networks are just a fraction of the achievements possible with the ability to harness technology.

As the world evolves, unknown challenges will present themselves to all nations, which must be well-equipped to handle changing needs and circumstances. Engineers must continue to strike at the heart of new challenges, simplifying problems to their fundamental roots to provide systematic, elegant solutions for advancing society through uncertain territory.

Singapore, in her first 50 years since independence, has advanced from an economy reliant on low labour cost to a highly developed, reputable global player renowned for reliability and high-value addition. Today, as the information-driven world shifts at rapid speed, Singapore can either isolate herself from the current of progress or embrace and evolve with the changes that disruption brings.

As PM Lee aptly put it during his 2016 National Day Rally speech, "Old models are not working, new models are coming thick and fast, and we're having to adjust and keep up". Indeed, if we choose to adapt, amidst the differing challenges that disruption brings to various industries, there exists a wealth of opportunities and value-addition to unearth if disruption can be tapped upon.

On the governmental level, the Committee on the Future Economy (CFE), EDB, SPRING Singapore and Infocomm Development Authority (IDA) have been working on strategies and practices to spot trends and help companies adopt new practices to ride on technology. The ultimate outcome of such efforts is to make our companies more resilient and our future workforce more nimble and well-prepared to meet a shifting economic outlook.

We will need a clear and effective plan that embraces a shared vision and commitment to the necessary restructuring and mindset changes. Engineers For the Future (EFF) aims to be a statement on how we can revitalise the national engineering enterprise, and mobilise technologies and resources to create a resilient modern economy based on technological strength. EFF is also an action plan to improve the perception and enhance the performance of engineers present and future.

To stay ahead of the competition, we need to develop a culture that is both entrepreneurial and adaptive, and that is holistically more inquiring. We need to acquire deeper skills to create value, and more importantly ensure that we can utilise our skills effectively on the job.[1] We need a mindset change that can address and reward efforts and determination to transform ideas into products and services of high quality and meeting global standards.

And the time has come for us to prepare for an aging population and leverage information and technology to enable us to live a good and healthy life. Furthermore, in this new millennium of low carbon future, our city state will require creative and elegant solutions to address the Grand Challenges on sustainability, and meet the need to mitigate and adapt to climate change.

Among the famous Grand Challenges of the 21st Century collated by the US National Academy of Engineering,[2] 6 stand out with stark relevance to Singapore. These are:

Grand Challenge 1: Make solar energy more affordable
Grand Challenge 5: Provide access to clean water
Grand Challenge 6: Restore and improve urban infrastructure
Grand Challenge 7: Advance health informatics
Grand Challenge 8: Engineer better medicines
Grand Challenge 13: Advance personalised learning

The above Grand Challenges map squarely those facing Singapore and nations in our region. In addition, Grand Challenge 14: Engineer the Tools of Scientific Discovery has particular relevance to transforming Singapore towards an economy that is driven by innovation and knowledge creation.

The millennium grand challenges were distilled from inputs solicited from prominent engineers and scientists and leading thinkers from around the world. No matter to which corners of the globe they may apply, the challenges await creative and elegant engineering solutions.

Singapore is dependent totally or to a large extent on imports to meet our domestic energy, food and water needs. The rising population, increasing demand for better quality of life, and the need to rein in carbon emissions will put a strain on our resources in the face of environmental challenges and rising costs.

To ensure sustainable economic growth and remain attractive to investment and inflow of talent, Singapore cannot compete in low cost production and traditional

---

[1]CFE, "Report of the Committee on the Future Economy: Pioneers of the next generation", Ministry of Communication and Information, February 2017, https://www.gov.sg/microsites/future-economy/the-cfe-report/read-the-full-report
[2]United States National Academy of Engineering, "NAE Grand Challenges for Engineering", http://www.engineeringchallenges.org/challenges.aspx

manufacturing businesses. At the same time, Singapore's economy is vulnerable to the influence of global economic shifts and geopolitical developments such as the developing China's One Belt One Road initiative and the new US outlook on world trade.

To build resilience and remain economically competitive, engineering will need to remain as a key enabler to a technologically driven global economy supported by well-trained professionals who are highly valued for their problem solving skills, and who can contribute to the growth of new industries.

We ask "how should we craft our own national grand challenges, and how do we prepare ourselves to meet them in this new millennium"? Engineering challenges unique to Singapore include climate change mitigation and adaption, green manufacturing technology, corrosion technology for hot and humid climate, engineering solutions to improve food security, low energy air conditioning systems, technology solutions for enhanced personal mobility, electrification of our transportation system, enhanced water and energy production, intelligent homes and smart meters, biofuels and emissions control technology, and low cost high yield humanitarian technologies and innovations.

We believe we have some of the answers but we recognise that there are gaps, which for the most part concern the formation and retention of talented engineers. To meet future challenges in sustainable living and economic development, we aim to develop young talent in engineering so that in time these engineers will be well-prepared to lead in technology and innovation. If not addressed, the gaps caused by shortfalls of talented engineers will pose a huge handicap in the economic development of the next 50 years.

We recognise that education underpins the core of a resilient society that can hold its own against competition and is adaptive in a changing world. Education threads through our schools and institutions of higher learning (IHLs) to continuing professional training. The formation of the future engineer can be a collective effort by the IHLs, IES and SAEng leveraging experiential learning, internships and mentoring. To attract the young to take up engineering, enthuse the enrolled engineering students and help retain career engineers, a holistic programme needs to be designed and delivered with focus on the following topics:

- Engineers as Leaders
- Sustainable Engineering
- Engineering for Societal Impact
- The Global Engineer
- Engineering Education for the Future
- Sparking Engineering Interest
- Engineering Contributions to National Development & Economic Competitiveness
- Engineering to Support Local Enterprises
- Developing Engineering Start-ups

We have an urgent need to move beyond being strong in service and maintenance to being able to productise ideas and pioneer frontier technologies.

Actions Needed

The IHLs are training thousands of engineering graduates every year. However, the lack of recognisable and rewarding career pathways is turning engineering graduates away from the engineering profession.

There is a lack of engineering jobs in research, development, and design of innovative products to attract talented engineers. Even at area where R&D jobs are available, there are great difficulties in finding suitable candidates. The imbalance in supply and demand and the mismatch in expectation and reality form ultimate impediments towards a knowledge-based economy.

Singapore needs a coordinated outreach programme targeting secondary schools and career counsellors, led by the engineering community. The programme should aim to improve awareness about engineering and what engineers do and to enthuse young people about engineering and engineering careers.

It is imperative that the young be advised to make the subject choices that will enable them to pursue engineering and further on to embark on a rewarding engineering career in the industry. This nationwide programme should link schools with local employers, giving students the opportunity to learn more about the world of engineering work. It should be designed to create the next generation of engineers, by increasing the number of young people choosing an engineering career through a well-structured and coordinated approach.

We need to provide on-the-ground support for employers and help them to improve the reach, quality and impact of their schools engineering outreach and careers inspiration activities. In addition, our strategy needs to reach out to the wider population in Singapore to communicate that studying science and mathematics subjects at schools and colleges can open up a whole range of exciting and rewarding career opportunities.

Specifically, the urgent needs are in the following areas:

- A pool of engineers enculturated with a global mindset, with an aptitude to innovate and competence to conceive, design, implement and operate.
- A fleet of engineering enterprises with capabilities in research and development of top-notch technologies, generating know-hows, and making prestigious global brands and products.
- An ecosystem for educating, training, continuously upgrading, attracting, and retaining sought-after individuals as engineers for the future in Singapore.

Index

3D printing, 81

A*STAR Computational Resource Centre (A*CRC), 148
A*Star, 68, 73, 144
Abbott, 73, 74
Abrasive Blasting Regulation, 153
accessibility, 91
Accident Prevention in Shipyards, 153, 154
accident rates, 152
Accredited Checkers (AC), 161
Active, Beautiful, Clean waters (ABC Waters) Programme, 33
Adult Education Board (AEB), 60
advanced engineering techniques, 164
advanced membrane technologies, 25
aerospace, 57
aerospace avionics, 120
aerospace electronics, 121
aerospace engineering, 103, 115
aerospace hub, 103, 119
aerospace machining, 120
aerospace technology, 120
Agency for Science, Technology and Research (A*STAR), 116, 140
air cargo, 110
air cargo distribution, 108
air cargo logistics, 120
air conditioning system, 11
air hub, 107
air quality, 11
airbus, 115
aircraft engineering, 120
aircraft management, 120
airline pilots, 120
Airport Logistics Park of Singapore (ALPS), 111
alternative energy, 50

alternative materials, 89
Anchor Handling Tug Supply (AHTS), 145
Ang Mo Kio Town Centre, 90
Ang, How Ghee, 154
AP Moller Maersk, 138
Applied Materials, 70
artificial intelligence, 81, 134
ASEAN, 46
ASEAN Power Grid, 46
Asia Pacific Brewery, 72
Assembly & Test (A&T), 67
assistive robotics, 133
Association of Aerospace Industries (Singapore) (AAIS), 117
authorised examiners, 151
automated equipment, 77
Automated Material Handling System (AMHS), 68
automated processing, 72
Automatic Guided Vehicle (AGV), 68
automatic irrigation system, 97
autonomous vehicles, 80, 133
average daily ridership, 2, 12
aviation management and services, 121
aviation skills training, 122
aviation, 103
Ayer Rajah Expressway (AYE), 18

Bachelor of Engineering, 61
Balakrishnan, Vivian, 133
barrier-free design, 90
BCA Academy, 98
BCA Green Mark, 92
BCA Green Mark Gold, 131
Bell Helicopters, 117
Bennett & Associates (BASS), 145
Bethlehem Singapore, 144
big data analysis, 81

bilateral agreements, 25
bioclimatic vegetation, 96
biodiversity, 17, 35
BioMatrix, 71
biomedical sciences, 57
Biopolis, 73
bioretention swales, 34
Boeing 747, 106
Boeing, 115, 122
boil-off gas (BOG) compressors, 48
Braille plates, 91
British Airways, 113
British Petroleum Company Limited, 75
broadband Internet access, 127
Broadband Media Association, 128
Building and Construction Authority (BCA), 90
Building Control Unit, 160
building envelop, 94
building greenery, 92
Building Information Modeling (BIM), 88
Building Management System (BMS), 97, 99
Building, 58
Bukit Timah Expressway (BKE), 2, 17
Bukit Timah, 15, 17
bulk tube handlers, 77
bus, 1, 12
bushfires, 26

CAD/CAM, 138
CAD/CAM technology, 138
Caltex Petroleum Corporation, 75
Capability Development Programme, 116
car ownership, 2
carbon dioxide, 47
carbon emissions, 12
carbon footprint, 39, 40, 52, 100
carbon-intensive, 48
Cargo ground handling, 109
cargo hubbing, 108
cargo logistics, 103
Caterpillar, 108
Central Business District (CBD), 17
Central Catchment nature reserves, 17
Central Expressway (CTE), 17
Centre of Innovation (COI) for Marine and Offshore Technology, 140
Centre of Innovation (COI), 140
certified inspectors, 157

challenges, 45
Changi Airfreight Centre (CAC), 111
Changi Airport Extension, 2
Changi Airport, 15, 103, 104
Changi East, 107
Changi International Airport Services (CIAS), 110
Changi Water Reclamation Plant, 32
chartered engineers, 151
chemical engineering, 152
chemical over-absorption, 159
Chew, Pin Kee, 153
Circle Line (CCL), 2, 13, 160
circle line stage 6, 15
city hall, 4, 13
civil and structural, 151
Civil Aviation Authority of Singapore, 121
civil engineering, 83
Claus Hemmingsen, 138
CleanTech Park, 52
Clementi, 4
climate change, 26, 37
Clinical Nutrition Research Centre (CNRC), 73
Closed Circuit Television (CCTV) System, 6, 9
cloud computing, 129
CO_2 emissions, 21
coal, 46
Colt, 127
Commission of Inquiry, 158
commissioning, 137
Committee of Inquiry, 160
committees, 155
communications, 5
community activities, 33
community involvement, 34
community spaces, 34
comprehensive traffic study, 2
comprehensive transport study, 3
compulsory health assessments, 155
computational fluid dynamics (CFD), 131
Concorde, 113
concrete, 89
concrete strengthening, 89
confined spaces, 158
Conoco platform project, 145
constant flow regulators, 26
construction industry, 85

Construction WSH Leadership Summit, 165
contactless smart cards, 8
contact-tracing, 160
control measures, 163
conversion projects, 136
conveyance channel, 35
cooling load, 92
Coolport @ Changi, 111
Corporate Lab, 118
cost-effective, 31
Cranfield University, 122
Creative Technology Limited, 64
Cross Island Line, 15
cyber-physical systems, 80
cybersecurity, 134

Data Storage Institute, 68
Dawson Estate, 101
daylighting, 93
death rates, 152
Debarment Scheme, 159
Deep Tunnel Sewerage System (DTSS), 14, 26, 31
deep tunnel sewers, 31
deep water operations-red, 138
Deepwater Technology Group (DTG), 140, 146
Defence Science and Technology Agency (DSTA), 123
Demerit Points System (DPS), 159
Department of Industrial Health (DIH), 157, 159
Department of Industrial Safety (DIS), 157, 158
Deputy Prime Minister, 4
desalinated water, 25
desalination plants, 25
desalination, 71
Design for Safety (DfS), 166
DfS experts, 166
dial-up, 127
digital sound card, 78
digital tools, 80
digitalisation, 133
diploma, 61
disc drives, 57
disk-drive technologies, 62
distribution centres, 111
Downtown Line Extension, 15

drains, 25
drilling rigs, 138
drillship, 138
dry dock, 135
dry wall, 89
DSO National Laboratories (DSO), 121, 123
DSS, 147
dual flush, 26
dust control measures, 153
Dynamic Compaction method, 106
dynamic positioning accommodation and repair vessel, 136
dynamic positioning system (DPS), 147

EADS (Airbus), 116
Eagle, 143
Eastern Region Line, 15
East–West Line (EWL), 4
Eco-link@BKE, 17
École Supérieure d'Ingénieurs en Électronique et Électrotechnique (ESIEE), 61
ecological balance, 17
Economic Development Board (EDB), 56, 57, 61, 71, 117, 140
eco-system, 79
education & training, 120
Edwards Lifesciences, 71
EIFS, 8
electric car, 52
Electrical and Energy (EE) engineers, 39
electrical, 152
electricity supply networks, 39
Electronic Private Automatic Exchange (EPAX) System, 5
Electronic Road Pricing (ERP), 15
Electronics and Electrical, 58
Embry-Riddle Aeronautical University, 122
emissions, 47
energy conservation, 10
energy consumption, 92
energy efficiency, 10, 92
Energy Management, 98
Energy Management System, SCADA, 45
Energy Market Authority, 45
energy security, 48
energy sources, 45
energy sustainability, 10, 40

energy-efficient, 131
energy-efficient lighting, 10, 96
enforcement, 155
Engineering Service Providers (ESPs), 72, 75
engineering, 1, 13
Enhanced Integrated Fare System (EIFS), 8
environmental contamination, 33
environmental impact, 12
environmental protection, 10
environmental protection, 92
equal employment opportunities, 161
equipment engineers, 67
ESEMI, 146
Esso, 74, 75
Eunos, 15
Execujet, 117
export-oriented industries, 56
Expressway Monitoring and Advisory System (EMAS), 15, 20, 21
Extended Semi-submersible, 146
ExxonMobil, 75
EZ-link cards, 8

façade greening, 98
Factories (Building Operations and Works of Engineering Construction) Regulations, 158
Factories (Crane Drivers and Operators) Regulations, 157
Factories (Medical Examinations) Regulations, 157
Factories (Qualifications and Training of Safety Officers), 155
Factories (Safety Committee) Regulations, 155
Factory Inspectorate, 155, 156
Faculty of Engineering, 61
fail-safes, 5
failure analysis engineers, 67
Far East Levingston Shipbuilding (FELS), 138, 144
Far East Levingston Shipyard (FELS), 143
fare card, 7
FAST (fast and seamless travel), 106
fatality rates, 162
fatigue cracking, 9
flash floods, 26

Floating Liquefied Natural Gas Vessel (FLNGV), 137, 149
Floating Production Storage Offloading (FPSO), 136, 145, 147
floating production storage and offloading (FPSO) vessel conversion, 135
Floating Storage and Offloading (FSO), 147
Floating Storage and Re-Gasification Unit, 137
Floating Storage and Regasification, 147
flood control, 31
Fokker Services Asia, 117
Food and Beverage, 58
Food Innovation and Resource Centre (FIRC), 73
Ford Motor Works, 55
foreign labourers, 158
formwork, 89
fossil fuels, 40
four national taps, 25
Fraser & Neave, 72
Free Trade Zone (FTZ), 110
freight forwarding, 109
French-Singapore Institute (FSI), 61
Friede & Goldman (F&G), 144
Front Opening Universal Pod (FOUP), 68
Furniture, 58
Fuyu Mould, 65, 66

Gas Technology Development (GTD), 149
GEC-Marconi, 127
general-purpose input/output (GPIO) utility, 131
geotechnical work, 161
German-Singapore Institute (GSI), 61
GlaxoSmithKline (GSK), 73
GMP, 73, 74
Goh, Chok Tong, 4
Goh, Keng Swee, 123
GPS, 19, 131
graving dock, 135
green buildings, 83, 91, 95
green building technologies, 97
green features, 92
Green Man Plus system, 21
Green Mark Platinum, 99
green roof, 97
green solutions, 52
greenery, 93, 95

GreenLite, 12
Guinness world record, 145

HACCP, 73, 74
hard disk drive industry, 68
Hawker Pacific Asia, 117
hazard control, 159
hazard identification, 153
Health and Safety Engineering (HSE), 151, 156
Heterogeneous network (HetNet), 129
Hewlett Packard, 64
High-Speed Rail, 21
high-tech industries, 57
high-value products manufacturing, 55
high-volume production, 77
Hitachi Zosen Robin Dockyard, 143
Hon Sui Sen, 142
Honeywell, 116
Hotel New World, 158
Housing & Development Act, 84
Housing & Development Board (HDB), 41, 84, 85, 97
HP Singapore, 64
human capital development, 120
hybrid electric vehicles (HEVs), 21
hydrogen fuel cell, 12
Hyflux, 71

ice-breakers, 138, 147, 148
Imperial Airways, 112
imported water, 25
incineration plants, 50
independence, 3
Industrial and Systems Engineering, 61
industrial engineers, 67
industrial estates, 56
Industrial Health Division (IHD), 155
Industrial Health Unit (IHU), 153
industrial hub, 152
Industrial Hygiene Monitoring Programme, 155
Industrial Postgraduate Program (IPP), 117
Industrial Training Board (ITB), 60
industrialisation, 55, 56, 152
Industry 4.0, 134
Infocomm Development Authority of Singapore (IDA), 128

Infocomm Media Development Authority (IMDA), 129
Infocomm Technology (ICT), 125, 128
information technology, 57, 159
Infrared Fever Screening System, 124
injection moulding, 66
innovation, 13
Institute of Technical Education (ITE), 120
Institute of Technical Education, 60
Institution of Chemical Engineers, 48
Insulated Gate Bipolar Transistor (IGBT) inverters, 12
integrated circuit (IC), 67, 77
Integrated Communication System (ICS), 4
integrated ocean drilling vessel, 136
Integrated Services Digital Network (ISDN), 126
Integrated Ticketing System (ITS), 7
Intel, 68
intelligent sensors, 19
Intelligent Transport Systems (ITS), 19
interactive information service, 127
Interactive services, 20
internal combustion engine vehicles (ICEVs), 21
International Advisory Panel (IAP), 165
International Civil Aviation Organisation (ICAO), 122
International Direct Dialing (IDD), 125
International Labour Organization (ILO), 152
international practices, 73, 74
International Social Security Association, 168
internet connectivity, 79
Internet of Things, 133
Internet penetration rate, 126
Internet service providers (ISPs), 126
inverter system, 100
Inverters, 10, 12
IOT technology, 80
iPartners Programme, 116
Ishikawajima-Harima Heavy Industries, 142

jack-up rigs, 135, 138, 147
Jalan Besar, 31

Japanese-Singapore Institute of Software Technology (JSIST), 61
Jet Aviation Singapore, 117
Jetstar Asia Airways, 114
Johor Bahru, 21
Johor, Malaysia, 25
JTC, 52
Jurong, 15
Jurong Industrial Estate, 56, 62, 135
Jurong Island, 74
Jurong Port, 56
Jurong Region Line, 15
Jurong Rock Cavern, 75
Jurong Shipyard, 142
Jurong Town Corporation (JTC), 85, 142

Kall Teck, 143
Kallang, 15
Kallang Airport, 111
Kallang Basin, 14
Kallang Marine Industrial, 135
Kallang Marine Industrial Estate, 135
Kallang River @ Bishan-Ang Mo Kio Park, 35
Kallang–Paya Lebar Expressway (KPE) tunnel, 14
Keppel FELS, 139, 145
Keppel O&M Technology Centre (KOMtech), 147
Keppel O&M, 138, 139, 144
Keppel Offshore & Marine (Keppel O&M), 136
Keppel Offshore and Marine Technology Centre (KOMTech), 139
Keppel Shipyard, 140
Keppel Singmarine, 145
Keppel Wharves, 143
Keppel's Floating Storage Regasification and Power (FSRP), 149
KFELS A Class, 145
KFELS B Class, 145
KFELS B Class Bigfoot, 146
KFELS Super B Class, 146
KFELS Super B Class Bigfoot, 146
Kim Chuan Telecommunications Centre 2 ("KCTC-2"), 130
knowledge intensive, 55
knowledge-based enterprises, 164
knowledge-based industries, 62, 70
knowledge-intensive industries, 60

Kranji Expressway (KJE), 15, 17
Kwong Soon Engineering, 143

labour intensive, 55, 62
labour-intensive industries, 60
land reclamation, 106
Land Transport Authority (LTA), 2, 10
Land Transportation Policy Division, 2
Land Transportation, 1
Land usage sustainability, 12
land use, 12
lean manufacturing, 69
Lee, Hsien Loong, 128, 133, 152
Lee, Kuan Yew, 4, 109, 142
Lee, Yi Shyan, 86
licensed aircraft engineers, 121
life science analytical technologies, 71
Life Technologies, 71
lifting equipment, 157
lifting supervisors, 157
Light Rapid Transit (LRT), 2
light-emitting diode (LED), 10
Lim Kong Puay, 47
Lim, Kim San, 84
Linear Variable Differential Transformer (LVDT), 10
Liquefied Natural Gas (LNG), 135
liquefied natural gas (LNG), 41, 45, 47
lithium-ion battery, 12
LNG bunkering vessel, 149
LNG carriers, 136
local catchment, 25
low-cost manufacturer, 55
low-emissivity coating, 98
Lynx, 127

M1, 127
Made-in-Singapore, 58
Maersk, 147
magnetic ticket, 7
Maintenance Omnibus, 6
Maintenance, Repair and Overhaul (MRO), 103, 114, 116
major hazards installations (MHIs), 166
Malayan Airways, 109
Malayan Airways Limited (MAL), 112
Malaysian Airlines, 109
Malaysia-Singapore Airlines (MSA), 109
Management Workshop on Safety in Shipyards, 154

Mandai, 17
manholes, 158
Manufacturing Excellence Award (MAXA), 145
Marathon LeTourneau, 143
Marina Barrage, 15, 26, 27, 30
Marina Barrage Solar Park, 41
Marina Channel, 26, 27
Marina Coastal Expressway (MCE), 15
Marina Reservoir, 26
Marine Technology Development (MTD), 147
Marine Technology Group, 140
Maritime and Port Authority of Singapore (MPA), 140
Mass Rapid Transit (MRT), 2–4, 7, 10, 13
Mass Rapid Transit Corporation (MRTC), 157
Massey University, 122
Matsushita Refrigeration Industries (S) Private Limited, 63
Maxtor, 68, 69
mechanical, 151
membrane, 27
metal stamping, 58, 66
Metalworking Safety Orientation Course, 164
MHI Regulations, 166
microfiltration, 27
Minister of Communications, 2, 3
Ministry of Defence (MINDEF), 123
Ministry of Education (MOE), 61
Ministry of Finance (MOF), 159
Ministry of Health (MOH), 153, 159
Ministry of Labour (MOL), 153
Ministry of Labour, 59
Ministry of Manpower (MOM), 159–161
MIT-Center for Environmental Sensing and Modeling (CENSAM), 35
Mitsubishi Singapore Heavy Industries, 143
Mobil, 74, 75
Mobile Earth Station (MES), 132
Modems, 132
modular heat exchange chillers, 100
monsoon season, 93
Motion sensors, 96
Mountbatten, 14
Mr Lee Kuan Yew, 23, 27

MRT Construction Hygiene Monitoring Programme, 156
MRT Corporation, 2
MRT system, 10
MRT tunnels, 157
multinational corporations (MNCs), 65
multiplexer, 132
municipal waste, 50
MyRepublic, 127

NAND Flash technology, 68
Nanyang Polytechnic, 61
Nanyang Technological University (NTU), 52, 121
National Environment Agency, 47
national fibre optics network, 127
National Industrial Safety and Health Campaign, 153
National Iron and Steel Mills, 62
National Library Building, 95
National Parks Board (NParks), 17
National Productivity Board (NPB), 59
National Productivity Council, 59
National Research Foundation, 118
National Safety First Council, 152
National Science and Technology Board, 144
National Solar Repository, 52
National Trades Union Congress, 152
National University Health System, 73
National University of Singapore (NUS), 52, 61, 121
National WSH Campaign 2013, 152
NatSteel, 62
natural lighting, 11
nature reserves, 17
NBAPs (Non-Building Access Points), 129
Neptune Orient Lines, 142
Nestle, 72
NEWater, 25, 27
Next Generation Nationwide Broadband Network (Next Gen NBN), 128, 129
Ng, Eng Hen, 161
Nicoll Highway, 160
Nicoll Highway stations, 14
nitrogen oxide, 47
North–East Line, 2, 4
North–East Line Extension, 15
North–South corridor, 18
North–South Expressway (NSE), 18

North–South Line (NSL), 3, 4
North–South Line Extension, 15
nose-to-tail capabilities, 117
Novena, 4
NTU, 12
NTUC, 161
NUS Centre for Offshore Research and Engineering (CORE), 140
NUS-EADS Internship Programme, 121

occupational diseases, 157
occupational health, 152
occupational safety, 151
Occupational Safety and Health, 159
Occupational Safety and Health Committee, 152
offshore and marine, 135
offshore and marine industry, 135
Offshore and Technology Development (OTD), 140, 144
offshore engineering, 136
offshore rigs, 138
Offshore Support Vessels, 145
Offshore Technology Development, 140
oil & gas, 74
oil and petrochemical, 158
oil industry, 74
oil refining and petrochemical industry, 166
oil rig manufacturing, 74
Ong Teng Cheong, 3
Ong, Pang Boon, 154
on-the-job attachments, 120
operation engineers, 67
Optical Connection Controller (OCC), 6
optical fibre, 5
Oriental Telephone and Electric Company (OTEC), 125
orientation, 92
Outram Park, 4
Outstanding Maritime R&D and Technology Award, 149

Pacific Internet, 128
Panasonic Refrigeration Devices Singapore Pte Ltd. (PRDS), 63
Panasonic, 67
Pan-Island Expressway (PIE), 2, 15
Parker Aerospace, 116
Pasir Panjang Power Station, 44

passive design strategy, 97
Paya Lebar Airport, 105, 109
pedestrian-friendly communities, 19
permit to work, 154
personal computers, 57
pharmaceutical and biotechnology manufacturing, 74
Philips, 64
photovoltaic, 98, 99
Pine, 127
Pinnacle @ Duxton, 85
Pioneer Status, 57
plant inspections, 155
plug-in HEVs, 21
pneumatic waste conveyance system, 100
population growth, 18
Port of Singapore Authority (PSA), 140, 143
Postgraduate School of Engineering, 62
power supply, 39, 42
Power System Control Centre (PSCC), 45, 46
Power System Operation Division (PSOD), 45
PPL Shipyard (PPL), 144
Pratt & Whitney, 115, 119
precast, 88, 101
precision engineering, 65, 66, 77
Precision Engineering Institute, 61
prefabricated, 100
prefabrication, 86, 87
pressure vessels, 157
preventive maintenance, 9
preventive maintenance management, 67
Prime Minister, 23
printed circuit boards, 57
process and maintenance engineers, 73
process engineers, 67
process industry, 72
Process Safety Award, 48
Productivity and Standards Board (PSB), 59
productivity, 59, 85, 87, 89
Productivity Movement, 59
Professional Engineer (Chemicals), 166
Professional Engineers (PE), 151, 158, 161
Professional Engineers Board, 166
Project Jewel, 106
Promenade, 14
Promet Shipbuilders, 144

PUB, 25, 26, 33, 35
Public Address (PA) System, 7
public transport, 1, 2
public transport network, 12
public transport nodes, 19
public transport system, 2, 3, 10
public transportation, 1
Public Utilities Board (PUB), 43
Public Works Department (PWD), 1
Pulau Ayer Chawan, 44, 74
Pulau Keppel, 143
Pulau Merlimau, 74
Pulau Seraya Power Station, 44
Punggol Eco-Town, 97

qualified person, 151, 160
Quality Control Circles, 59, 67
quality of service (QoS), 130

Radio Communication Systems, 6
radio-access technologies (RATs), 130
Raffles Place, 4, 13
railway, 1
railway communications, 4
rain gardens, 34
rain sensors, 97
rainwater harvesting, 94, 100
rapid motorisation, 18
real-time information, 20
reclaimed water, 25
recreational activities, 31
recreational space, 35
recycled water, 11
refining, 74
refresher courses, 158
regenerative braking, 12
regenerative energy, 10
Regional Training Centre of Excellence, 122
Registered Safety Officer (RSO), 160
Registry of Vehicles, 2
regulatory framework, 152
reinforcement work, 89
reliability, 8, 10
renewable energy, 46, 48, 50
Republic of Singapore Air Force (RSAF), 115, 123
research and development (R&D), 116, 138, 139

Research and Development Assistance Scheme, 144
reservoirs, 25
reverse osmosis, 27
rig building, 136
risk management, 160, 162
rivers, 25
Road and Transportation Division, 2
road management, 15
robots, 90
Rollei, 60, 68
rolling stock, 11
Rolls-Royce, 115, 119
Rolls-Royce @ Nanyang Technological University, 118
routine inspections, 155
Royal Air Force (RAF) Seletar, 123
rubber and plastic, 58
rule of law, 104

safe work practices, 155
safety and health, 164
safety auditors, 158
safety awareness, 154
Safety Case Regime, 166
safety champions, 152
Safety Consultancy Group, 154
safety engineering, 156
safety instruction courses, 157
safety management program, 158
safety management system, 159
Safety Officer Training Course (SOTC), 153
safety orientation, 154, 155
Safety Orientation Courses (SOC), 158
safety standards, 162
safety supervisors, 158
safety, 48
Sand and Granite Quarries Regulations, 153
sand, 89
satellite broadband communications, 133
satellite communications, 132
satellite telemetry, 131
SATS Ltd (SATS), 109
School of Mechanical and Aerospace Engineering, 121
School of Mechanical Engineering, 121
School of Media and Info-Communications Technology (SMIT), 61

Seagate, 69
seismic research vessel, 136
Seletar Aerospace Park (SAP), 115, 117
Seletar Airbase, 107
Seletar Airport, 123
Seletar Expressway (SLE), 17
self-closing delayed action taps, 26
Semakau Landfill, 50
Sembawang Shipyard, 142
Sembcorp Marine, 136, 138, 139
Sembcorp Marine Technology, 139
semiconductor, 57, 58, 62, 67
semi-submersible platforms, 135
semi-submersible rigs, 138, 146
Senoko, 50
Series, 147
services, 120
Setron Limited, 63
Severe Acute Respiratory Syndrome (SARS), 159
sewerage, 37
SGS, 67, 68
shading devices, 95
ship repair, 135, 136, 142
shipbuilding, 135, 138, 142
shipbuilding and repair, 154
Shipyard Safety Instruction Course, 164
shipyard, 135, 136
SIA Cargo, 106, 113
SIA Engineering Company (SIAEC), 115, 116, 122
silicate dust, 153
silicon wafer, 56
silicosis, 152
SilkAir, 113
Siloso Beach Resort, 100
SIM University (UniSIM), 121
Sim, Wong Hoo, 64, 78
Sing Koon Seng, 143
Singapore Aerospace Maintenance Company, 123
Singapore Aircargo Agents Association (SAAA), 110
Singapore Aircraft Maintenance Company (SAMCO), 115
Singapore Airlines (SIA), 106, 109, 112, 113
Singapore Airlines, 105, 111
Singapore Aviation Academy (SAA), 121

Singapore Baggage Transport Agency Pte Ltd, 110
Singapore Cable Vision, 128
Singapore Improvement Trust (SIT), 84
Singapore Institute for Clinical Sciences, 73
Singapore Institute of Standards and Industrial Research (SISIR), 59
Singapore International Maritime Awards, 149
Singapore Manufacturers' Association (SMA), 58
Singapore Manufacturing Federation (SMF), 58
Singapore National Employers Federation (SNEF), 161
Singapore ONE, 127
Singapore Petroleum & Chemical Company (Private) Limited, 74
Singapore Polytechnic, 61
Singapore Refining Company Pte. Ltd. (SRC), 75
Singapore River, 23
Singapore Slipway, 143
Singapore Technologies Aerospace, 115
Singapore Technologies Engineering, 68
Singapore Telecom, 127
Singapore Telephone Board (STB), 125
Singapore Vocational Institute (SVI), 60
SingNet, 127
SingTel, 127
six sigma, 69
skill intensive workforce, 55
skills-intensive, 62
Sky gardens, 95
Skyhawk, 123
Skytrain, 106
Skyville @ Dawson, 101
small and medium enterprises (SMEs), 65, 72, 116
smart home solutions, 133
smart nation initiative, 46
Smart Nation, 133
Smart Water Grid, 35, 36
soft soil conditions, 14
Software Programmable Radio, 133
soil conditions, 83
soil erosion, 34
soil investigation, 161
solar cooling technology, 52

Solar Energy Research Institute of Singapore (SERIS), 52
solar energy, 41
solar glare, 95
solar heat gain, 92
solar panels, 52
solar photovoltaic (PV), 46, 48
solar photovoltaic panels, 41
solar radiation, 48, 92
solar technology, 41
solar wafer, 70
Sony Precision Engineering Center (Singapore), 67
Sony Singapore, 63
sound blaster, 64
ST Aerospace, 104, 115
ST Electronics, 4, 20
ST Engineering, 131
StarHub, 127
Start-up Enterprise Development Scheme (SEEDS), 57
State and City Planning (SCP) project, 1
station control room (SCR), 6
statutory board, 59
steel construction, 89
STMicroelectronics, 56, 67
stop work orders, 156
stormwater collection ponds, 25
stormwater management, 34
structural design, 158
submarine cables, 44
substations, 45
sulphur dioxide, 47
sun breakers, 101
sunrise industries, 57, 156
sunshade, 95
Super Group, 72
supersonic jet, 113
Supervisory Control and Data Acquisition System, 45
sustainability, 26
sustainable development, 52, 91
sustainable technology and design, 39
sustainable water supply, 23, 25
Suzhou-Singapore Industrial Park, 57

T5 fluorescent technology, 10
Tailored Support Programme, 115
Tampines Expressway (TPE), 17
Tampines, 15

Tan Henn, 78
Tanjong Pagar container terminal, 143
Tanjong Rhu, 143
Tanjong Rhu Basin, 135
Teamy the Productivity Bee, 59
technical support, 120
technical training, 60
technical workers, 55
Technische Universitat Munchen (or the Technical University of Munich), 52
technological innovations, 10
technology intensive, 55, 62
Telecommunication Authority of Singapore (TAS), 125
telecommunication, 125, 127, 131
Telematics Services Hub/oTTo-Go, 20
Telephone Department, 125
Teleview, 127, 157
Temasek Polytechnic-Lufthansa Technical Training Centre, 121
Temporary Earth Retaining System (TERS), 161
Terminal 1, 105, 106
Terminal 2, 105
Terminal 4, 106
Terminals 3, 104, 105
Texas Instruments, 67
Textile, 58
Thales, 116
The Institution of Engineers, Singapore (IES), 128
Thermal analysis, 131
third rail, 10
Thomson Line, 15
Thomson Road, 18
Thong Siek Food Industry, 72
Tiger Airways, 114
Tiong Bahru, 4
Toa Payoh, 3, 15
Top 50 Engineering Feats, 128
top–down construction, 14
topsides, 138
track hump profile, 12
Trackside Emergency Trip Station, 6
trade routes, 104
Tradewinds Airlines, 113
traffic conditions, 20
Traffic management, 19
train disruptions, 9
trainee engineers, 121

Transit-oriented development (TOD), 19
transmission and distribution networks, 39, 41
Transmission System, 5
transport, 1
Treelodge @ Punggol, 97
Tsinghua University, 12
Tuas Biomedical Park (TBP), 73
Tuas Power Ltd, 47
Tuas South, 50
Tuas Water Reclamation Plant (WRP), 36
Tuas West Extension, 15
Tuas, 15
Tuas, 50
TUM CREATE, 52
tunnel engineers, 158
tunnelling works, 156
TV assembly, 63

ultra-high speed broadband, 129
ultraviolet disinfection, 25, 27
underground construction, 31
underground MRT network, 13
Underground Road — Marina Coastal Expressway (MCE), 15
underground, 83
underground seepage, 31
University of California, Berkeley, 122
University of Malaya (UM), 61
University of Singapore, 61
unsafe conditions, 152
urban farms, 95
urban heat island, 97
urban planning, 18
USB thumb-drive, 78

variable frequency drives, 48
variable speed drives, 99
Vertical Drain methods, 106
vertical greening, 98
video-streaming, 128
Vietnam-Singapore Industrial Park, 57
ViewQwest, 127
Visenti Pte Ltd, 35
Vision Zero, 165
Vocational and Industrial Training Board (VITB), 60
voice synthesiser, 91

wafer fabrication, 67
waste heat recovery, 37
waste-to-energy, 50
water conservation, 25
water demand, 25
water distribution, 37
water efficiency, 92
water engineering, 37
water infrastructures, 32
water leakages, 31
water pollution, 23
water reclamation plant, 26
water recycling, 32
water security, 23
water supply, 26
water sustainability, 25
Water-saving devices, 26
Weng Chan, 143
Western Digital, 68, 69
wetlands, 34
Wheel Impact Load Detection (WILD), 10
Wireless @ SG, 128
wireless network, 80
WIT, 67
WOHA, 101
Woodlands, 17
Work Improvement Teams, 59
work permits, 158
Worker Safety and Wellbeing Test, 164
Workforce Development Agency (WDA), 164
workmen compensation, 151
workplace hazards, 163
Workplace Safety and Health (WSH), 151, 155, 159
Workplace Safety and Health Council, 162
Works of Engineering Construction Regulation, 153
WSH (Design for Safety) Regulations, 166
WSH (Risk Management) Regulations, 163
WSH Act, 162
WSH Action Plan, 166
WSH Advisory Committee, 162
WSH culture, 162, 164
WSH framework, 160, 161
WSH Institute, 164

WSH Professionals Workforce Skills
    Qualifications, 164
WSH standards, 155, 159

X-Sat, 121
XXI World Congress on Safety and
    Health at Work, 168

Yeo, Philip, 57
Yio Chu Kang, 3

Zero-Energy Building (ZEB), 98

www.ingramcontent.com/pod-product-compliance
Lightning Source LLC
Chambersburg PA
CBHW080322170426

43193CB00017B/2880